PRAISE FOR

The French and Indian War

"A popular military account of the war's campaigns and battles that prunes back on detail. . . . Introducing the war's prominent commanders, from Edward Braddock to Montcalm to Pontiac, Borneman keeps a respectful eye on the war's bloody cost as he fluently acquaints readers with its strategic course." —*Booklist*

"Historical writer Borneman . . . has produced a fast-moving popular history of the war. . . . Provides a rich narrative of the important campaigns, battles, and personalities."

—*Library Journal*

"Excellent. . . . Drawing on a broad spectrum of primary and secondary sources, Borneman . . . argues that the French and Indian War not only made Britain master of North America but created an empire that dominated the world for two centuries."

—*Publishers Weekly*

"Evinces much reading and a thorough understanding of the people, the places (many of which he visited), and the events. Evident, too, is a sort of narrative ebullience often lacking from more academic accounts. . . . Borneman is at his best elucidating battle strategies (especially the pivotal encounter at Quebec) and bringing to life the personalities of some of those famous names . . . who strode that particular stage." —*Kirkus Reviews*

"[*The French and Indian War: Deciding the Fate of North America*] by independent historian Walter R. Borneman, is a fast-paced introduction. . . . As its subtitle suggests, it demonstrates just how important the war was in configuring the world we inhabit today."   —*New York Times Book Review*

"It's in [Borneman's] step-by-step descriptions of the campaigns . . . that he'll rivet the attention of most readers. Borneman is a vigorous, passionate historian, and it's clear from the first that every battle, every general, heroic or inept, looms large in his understanding."   —*Providence Journal*

"Author Walter R. Borneman tells the story of the French and Indian War with diligence and enthusiasm. . . . Fresh and exciting throughout, *The French and Indian War* is highly recommended."   —BookLoons.com

The French and Indian War

ALSO BY WALTER R. BORNEMAN

14,000 Feet: A Celebration of Colorado's Highest Mountains
(with Todd Caudle)

1812: The War That Forged a Nation

Alaska: Saga of a Bold Land

A Climbing Guide to Colorado's Fourteeners
(with Lyndon J. Lampert)

The French and Indian War

DECIDING THE FATE OF NORTH AMERICA

WALTER R. BORNEMAN

HARPER ● PERENNIAL

NEW YORK ● LONDON ● TORONTO ● SYDNEY

HARPER ⬤ PERENNIAL

FIRST HARPER PERENNIAL EDITION PUBLISHED 2007.

Designed by Nancy B. Field

Maps by David Lambert

The Library of Congress has catalogued the hardcover edition as follows:

Borneman, Walter R.
  The French and Indian War : deciding the fate of North America / Walter R. Borneman.—1st ed.
    p. cm.
  Includes bibliographic references and index.
  ISBN: 978-0-06-076184-4
  ISBN-10: 0-06-076184-9
  1. United States—History—French and Indian War, 1755–1763.
  2. United States—History—French and Indian War, 1755–1763—Influence. 3.Canada—History—1755–1763. I. Title.

E199.B67   2006
973.2'6—dc22   2005058450

ISBN: 978-0-06-076185-1 (pbk.)
ISBN-10: 0-06-076185-7 (pbk.)

20 ❖/LSC 20 19 18 17 16 15 14 13

For my grandparents,
Walter and Hazel Borneman

CONTENTS

Maps

ACKNOWLEDGMENTS

My inclination after writing *1812: The War That Forged a Nation* was to look westward—both geographically and chronologically. Yet as I pondered my next project, I found myself drawn to events a generation *before* the American Revolution with the same fascination that I had just written about events a generation after it. Here was a period that decided the fate of the entire North American continent—not just between England and France, but among the Spanish and Native Americans as well. My goal became to present the triumphs and tragedies of this struggle; place them in the context of France and Great Britain's greater global conflict; essentially the first truly world war; and emphasize that from seeds of discord sown here grew the American Revolution.

With another book in hand, my high esteem and great appreciation only deepen for my editor, Hugh Van Dusen; and my agent, Alexander Hoyt. On the research side, it is always a pleasure to work in the Penrose Library of the University of Denver, and I must also thank the Van Pelt Library of the University of Pennsylvania, the Norlin Library of the University of Colorado, and the Denver Public Library. Additionally, I greatly appreciate the research assistance of Fadra Whyte at the University of Pennsylvania and Christopher Fleitas at the University of Notre Dame. David Lambert at National Geographic Maps contributed his cartographic skills.

In addition to colonial newspapers—which sometimes must be taken with a grain of salt—many primary sources from this period are increasingly available in published form. These include the personal papers and correspondence of such key figures as Amherst, Bougainville, Bouquet, Forbes, Franklin, Johnson, Pitt, Shirley, Wolfe, and of course the young George Washington. In quoting from contemporary accounts, I have taken the liberty to edit spelling, grammar, and capitalization, thereby avoiding the ubiquitous use of *sic*.

James Fenimore Cooper and Kenneth Roberts aside, there have been many scholarly histories of the French and Indian War over the years. Despite their heavy Anglophile biases, Francis Parkman's *Montcalm and Wolfe* remains a reference point and Lawrence Henry Gipson's epic fourteen-volume account of *The British Empire Before the American Revolution* an essential building block. To these long-established icons must be added Fred Anderson's recent *Crucible of War*, the most informative and best-written one-volume study of the period.

Other valuable secondary sources include Guy Frégault's *Canada: The War of the Conquest*, telling the story from the Canadian perspective; and Francis Jennings's *Empire of Fortune*, emphasizing the roles of Native Americans. More recent studies of Native Americans' interaction include Timothy Shannon's *Indians and Colonists at the Crossroads of Empire*, Tom Hatley's *The Dividing Paths*, and Matthew Ward's *Breaking the Backcountry*. For assistance in placing the North American campaigns in a global context, I found Walter L. Dorn's *Competition for Empire* and Alfred Thayer Mahan's *The Influence of Sea Power upon History* to have stood the test of time.

My favorite part of writing remains walking the ground where these events took place. Thus, my wife, Marlene, and I traveled Braddock's road, shivered in a cold wind on the ramparts at Fort Ticonderoga, sought out Rogers Rock, and pondered Pitt's moves in the Caribbean. Where to next, Marlene?

CHRONOLOGY

Key Dates of the French and Indian War

1753

December 11	Washington arrives at Fort Le Boeuf

1754

July 4	Washington surrenders Fort Necessity
July 11	Albany Congress adjourns
September 15	Braddock commissioned commander in chief

1755

February 20	Braddock arrives in Virginia
June 8	Boscawen captures the *Alcide* and the *Lys*
July 9	Braddock's Defeat on the Monongahela
September 8	Battle of Lake George

1756

May 18	Great Britain declares war on France
May 20	French defeat Admiral Byng off Minorca
June 9	France reciprocates and declares war on Great Britain
August 14	Surrender of British forts at Oswego

1757

June 29	Pitt-Newcastle ministry takes office
July 25	Duke of Cumberland defeated at Hastenbeck
August 5	Lord Loudoun abandons attack on Louisbourg
August 9	British surrender Fort William Henry
November 5	Frederick the Great defeats French at Rossbach

1758

June 8	Wolfe leads British troops ashore in Garabus Bay
July 8	Battle of Fort Carillon (Ticonderoga)
July 27	French surrender Louisbourg to Amherst
August 27	Bradstreet raids Fort Frontenac
September 14	Grant's battle outside Fort Duquesne
October 12	Battle of Fort Ligonier
November 24	French abandon Fort Duquesne
December 29	British capture Gorée in Senegal

1759

January 16	Barrington lands on Martinique
January 23	Barrington lands on Guadeloupe
May 2	French surrender Guadeloupe to British
June 26	Wolfe's troops land near Quebec
July 25	French surrender Fort Niagara
July 26	French abandon Fort Carillon
July 31	French abandon Fort Saint Frédéric
August 18–19	Boscawen defeats de la Clue off Lagos
September 13	Battle of the Plains of Abraham (Quebec)
September 17	French surrender Quebec
October 4	Rogers and his rangers attack Saint Francis
November 20	Hawke defeats Conflans in Quiberon Bay

1760

April 28	Second Battle of Quebec (Sainte-Foy)
May 15	British relief fleet arrives at Quebec
August 8	Fort Loudoun surrenders to Cherokee
September 8	French surrender Montreal and all of Canada
September 13	Rogers departs Montreal for Detroit
October 25	King George II dies

1761

August 15	Bourbon Family Compact signed
October 4	William Pitt resigns his office

1762

January 4	Great Britain declares war on Spain
February 13	British capture Martinique
August 11	British capture Havana
October 6	British capture Manila in Philippines

1763

February 10	Treaty of Paris signed
May 7	Pontiac first tries to take Detroit
June 2	Chippewa capture Fort Michilimackinac
July 31	Battle of Bloody Run near Detroit
August 5–6	Battle of Bushy Run near Pittsburgh
August 10	Relief of Fort Pitt by British forces
October 7	Proclamation of 1763 issued by George III
November 16	Gage succeeds Amherst as commander in chief

IMPORTANT PERSONALITIES
OF THE
FRENCH AND INDIAN WAR

In the courts of Europe

George II, king of England, r. 1727–1760
George III, king of England, r. 1760–1820
Louis XV, king of France, r. 1715–1774
Philip V, king of Spain, r. 1700–1746
Ferdinand VI, king of Spain, r. 1746–1759
Charles III, king of Spain, r. 1759–1788
Maria Theresa, empress of Austria, r. 1740–1780
Frederick II (the Great), king of Prussia, r. 1740–1786
Elizabeth, empress of Russia, r. 1741–1762
Catherine II (the Great), empress of Russia, r. 1762–1796

Duke of Newcastle, British prime minister, 1754–1757
William Pitt, British prime minister, 1757–1761
Earl of Bute, British prime minister, 1761–1763
Lord Anson, first lord of the Admiralty
John Ligonier, commander in chief, British army
Marquise de Pompadour, mistress of Louis XV
Duc de Choiseuil, French minister of foreign affairs

Native Americans in North America

Attakullakulla (Little Carpenter), Cherokee chief in Carolinas
Hendrick, Mohawk chief killed at battle of Lake George
Oconostota, Cherokee chief of Overhill clans
Neolin, Delaware prophet, inspired revolt against non-Indians
Pontiac, Ottawa chief at siege of Detroit
Tanaghrisson (the Half King), Seneca chief who aided Washington
Teedyuscung, Delaware leader involved with treaty of Easton

Commanders in chief of British forces in North America

Edward Braddock, 1754–1755
William Shirley, 1755–1756
Lord Loudoun, 1756–1758
James Abercromby, 1758
Jeffery Amherst, 1758–1763
Thomas Gage, 1763–1775

In the British colonies in North America

James De Lancey, lieutenant governor of New York
James Glen, governor of South Carolina
Robert Dinwiddie, lieutenant governor of Virginia
William Shirley, governor of Massachusetts

Principal British commanders in the field and on the seas

Edward Boscawen, naval operations off Louisbourg and France
Henry Bouquet, Fort Duquesne and Indian campaigns
John Bradstreet, bateaux expert, Fort Frontenac raid
John Forbes, Fort Duquesne campaign and Indian relations
Lord Howe, Abercromby's deputy, killed in attack on Carillon
William Johnson, superintendent of Indian Affairs
James Murray, one of Wolfe's brigadiers at Quebec, attacked Montreal

Robert Rogers, famous as ranger; doomed to other failures

Charles Townshend, one of Wolfe's brigadiers at Quebec; politician

James Wolfe, Louisbourg and Quebec campaigns

In the French colonies in North America

Marquis de Duquesne, governor-general of New France, 1752–1755

Marquis de Vaudreuil, governor-general of New France, 1755–1760

Principal French commanders in the field and on the seas

Maximin de Bompar, naval operations in Caribbean and off France

Baron de Dieskau, commanded forces at Battle of Lake George

Comte de Bougainville, Montcalm's trusted aide-de-camp

Marquis de Galissoniere, admiral opposing Byng off Minorca

François-Gaston de Lévis, succeeded Montcalm and defended Montreal

François-Marie le Marchand de Lignery, commandant at Fort Duquesne

Jacques Legardeur de Saint-Pierre, commandant at Fort Le Boeuf

Marquis de Montcalm, commander in chief of French forces, 1756–1759

Pierre Pouchet, commandant at Fort Niagara

INTRODUCTION

THE WAR THAT WON
A CONTINENT

England and France had been at war since—well, it seemed like forever. For more than three centuries, Europe had known far more years of warfare than of peace. But no matter what the conflict, or how causes and alliances changed, one pairing remained constant: England and France were always on opposite sides just as surely as they sat on opposite sides of the English Channel. By the mid-eighteenth century, however, this cross-Channel feud began to take on major global dimensions, as it became evident that far more than the mastery of Europe was at stake.

The colonies that half a dozen nations had established in the New World were flourishing. By 1733, thirteen English colonies stretched along the Atlantic coast. But this territory was minuscule compared with French outposts and settlements that embraced half a continent—from the mouth of the Saint Lawrence River, westward across the Great Lakes, and down the Mississippi River to the Gulf of Mexico.

Spain, too, was a major player in North America, claiming Florida, Texas, and the headwaters of the Rio Grande as the northern fringes of its domain. In earlier European wars, North America had been mostly a sideshow; but by 1748, these English,

French, and Spanish empires were colliding in North America along ever-expanding frontiers. The bad blood of centuries-old European feuds was about to be spilled here as well.

Hemmed in by French claims, the English colonies squeezed between the Appalachians and the Atlantic coast grew uneasy. Virginia dispatched a twenty-one-year-old surveyor named George Washington west to tell the French on the upper Ohio River that they were trespassers. The French were cordial, but emphatic in their denial. When a British force under General Edward Braddock marched to the forks of the Ohio two years later, it met with a disastrous defeat that unleashed what quickly became history's first global war.

From the Ohio River to the falls of Niagara, across Lake Champlain, and down the Saint Lawrence River, North America's colonial frontiers erupted in flames. By the time what Europe called the Seven Years' War was concluded, it had been fought not only in North America, but also on the battlefields of Europe and in colonies throughout the world—from the Caribbean to India, Africa, and the Philippines.

The war in North America was characterized by desperate battles in virgin wilderness. There were epic treks by Rogers' Rangers, the original Green Berets; dogged campaigns to capture strategic linchpins such as Fort Duquesne and Fort Ticonderoga; and the legendary battle of Quebec atop the Plains of Abraham. Then, just when the British thought that they had won a continent, France counterattacked and almost recaptured Quebec.

When the warring powers finally met to sign the Treaty of Paris of 1763, the map of the world looked quite different from its appearance seven years before. As the historian Francis Parkman succinctly put it, "half a continent changed hands at the scratch of a pen." But a challenge soon came from France's Native American allies. Urged on by an Ottawa chief named Pontiac, a loose confederation of Northwest Indian nations launched a series of attacks that again turned the colonial frontier red with blood and threatened to lose for Great Britain all that it had gained from France.

Great Britain's resolution of this Native American resistance had almost as much to say about the future of North America as did its victories over France. King George III proclaimed a vast "Indian reserve" between the Appalachians and the Mississippi River, effectively hemming in his American colonists along the Atlantic coast just as the French had done previously. The young king also looked to the colonies as a source of income to pay the debts of this latest war.

Land claims extinguished west of the Appalachians and taxes imposed without representation quickly rankled colonists no longer bound to the British crown by the fear of French encirclement. Revolution was premature, but the die had been cast. The triumphs of one war had sown the seeds of discontent that would lead to another. Great Britain had indeed won a continent, but in doing so, it had also lit the fuse of revolution.

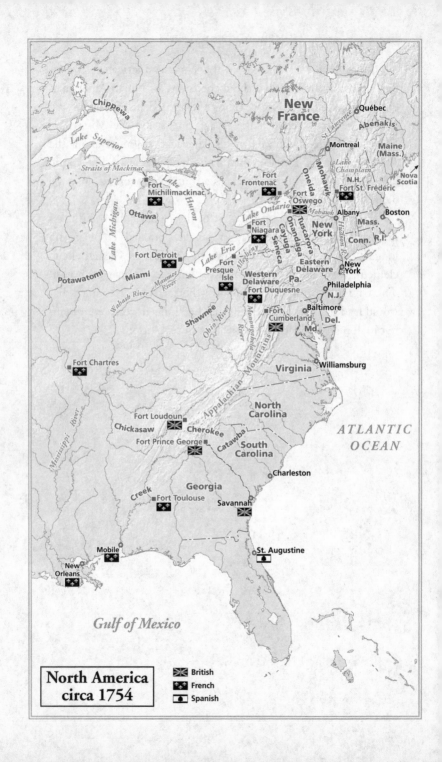

Chippewa

Lake Superior

New France

Québec

Abenakis

Montreal

Maine (Mass.)

Straits of Mackinac

Lake Huron

Fort Michilimackinac

Fort Frontenac

Lake Champlain

N.H.

Fort St. Frédéric

Nova Scotia

Ottawa

Lake Ontario

Fort Oswego

Fort Niagara

Mohawk

Oneida

Onondaga

Cayuga

Tuscarora

Seneca

New York

Albany

Mohawk

Boston

Mass.

Conn.

R.I.

Lake Michigan

Fort Detroit

Lake Erie

Fort Présque Isle

Allegheny R.

Eastern Delaware

New York

Potawatomi

Miami

Maumee River

Western Delaware

Pa.

Philadelphia

Hudson Rv.

Wabash River

Fort Duquesne

N.J.

Shawnee

Ohio River

Monongahela River

Fort Cumberland

Baltimore

Md.

Del.

Fort Chartres

Virginia

Williamsburg

Appalachian Mountains

Mississippi River

Fort Loudoun

North Carolina

Chickasaw

Cherokee

Fort Prince George

Catawba

South Carolina

ATLANTIC OCEAN

Georgia

Charleston

Creek

Fort Toulouse

Savannah

Mobile

St. Augustine

New Orleans

Gulf of Mexico

North America circa 1754

British

French

Spanish

BOOK ONE

Colliding Empires

(1748–1756)

For forming this general union, gentlemen, there is no
time to be lost; the French seem to have advanced further
towards making themselves masters of this continent
within these last five or six years than they have done ever
since the first beginning of their
settlements upon it.

—WILLIAM SHIRLEY, ROYAL GOVERNOR OF MASSACHUSETTS,
to the General Court of Massachusetts, April 2, 1754

1

THE BELLS OF
AIX-LA-CHAPELLE

In the fall of 1748, the bells in the venerable cathedral at Aix-la-Chapelle pealed out the welcome news that Europe was again at peace. Europe's warring powers had gathered at the site of Charlemagne's medieval capital in yet another attempt to end their incessant feuds and bring a lasting peace to the continent. But the Treaty of Aix-la-Chapelle—as has been the case with so many of history's paper pronouncements—failed to resolve gnawing geopolitical realities. The playing field was no longer just the continent of Europe. Increasingly, it was the entire world, and the rival empires that fought on European battlefields were also colliding on far-flung oceans and faraway continents. Nowhere was this truer than in North America.

For more than two centuries, England had lagged far behind its European rivals in coloring in the map of North America. John Cabot sailed the North Atlantic a few years after Columbus's first voyage, but England did little more for decades. Meanwhile, the Frenchman Jacques Cartier circled Newfoundland and probed the Saint Lawrence River as far as the site of Montreal in 1534–1535. Elsewhere, Coronado carried the Spanish banner across the American southwest and de Soto traversed the Deep South between 1539 and 1542. Somewhat belatedly, Elizabeth I of England sent Martin

Frobisher on three voyages across the Atlantic in the 1570s to search for the Northwest Passage and reassert Cabot's claims. By then, Spain had already established an outpost at Saint Augustine on the Florida coast in response to French forays into the area.

The continent of North America was never, of course, a universal blank waiting to be claimed, as all Europe deemed it. Numerous Native American nations, some quite powerful empires themselves, held sway over forest, lake, bayou, and river. Like the Europeans, they, too, frequently fought among themselves for territorial rights and other prerogatives. These Native American—or Indian—sovereignties did not deter European incursions, but they certainly made such incursions far more complex and conflicted. (Because contemporary accounts use the term "Indians," it has sometimes been retained here rather than the currently accepted term "Native Americans.")

In 1577, Spain was still the world's leading power, but Elizabeth could not resist probing its weaknesses by dispatching Francis Drake to circle the globe and cause a little havoc on the Spanish Main. Attempts by the English to plant a colony at Roanoke on the Carolina coast withered under the distraction of the Spanish Armada; and by the time three English ships anchored off Jamestown, Virginia, in 1607, Elizabeth was dead and England still far behind in the race for a continent.

Three years before, a French company that included Samuel de Champlain had established an outpost at the mouth of the Saint Croix River between present-day Maine and New Brunswick. After a damp and frigid northeastern winter, the post was moved to Port Royal on the northwest coast of Nova Scotia. When Port Royal was temporarily abandoned in 1607, Champlain and a few others refused to return to France and instead followed Cartier's route up the Saint Lawrence and built an outpost under the Rock of Quebec.

Meanwhile, Spanish Florida was thriving despite Drake's burning of Saint Augustine in 1586. Juan de Oñate's efforts at colonization on the northern reaches of Spain's claims resulted in the founding of Santa Fe, New Mexico, in 1609. That same year, the Dutch entered the competition when Henry Hudson, an Englishman serving under the Dutch flag, sailed the tiny *Half Moon* up the Hudson River. In 1624, just four years after the Pilgrims landed at Plymouth Rock, the Dutch West India Company built a trading post called Fort Orange at the future site of Albany, New York.

Thus the seeds of many would-be empires were planted in North America. For a time the vastness of the continent swallowed their meager expansion and prevented major collisions. Then in 1682, Robert La Salle and his men dragged canoes across the portage between the Chicago and Illinois rivers and floated down the Illinois to the Mississippi. Continuing south down the Mississippi, La Salle reached the Gulf of Mexico and on a spot of dry ground at the river's mouth proclaimed the sovereignty of Louis XIV over half a continent. The French already controlled one of North America's strategic arteries—the Saint Lawrence River and the Great Lakes—and now, on April 9, 1682, La Salle grandly claimed another. Henceforth, La Salle asserted, the Mississippi River and its tributaries, "this country of Louisiana," were the domain of France.[1]

Indeed, by 1700 a look at the map of North America suggested that France held claim to the lion's share. From Quebec, up the Saint Lawrence, across the Great Lakes to Michilimackinac, and down the Mississippi Valley, France constructed a string of trading posts that included Kaskaskia, Cahokia, and Vincennes in the Illinois country and Fort Pontchartrain (Detroit) between lakes Huron and Erie. Spain was equally well established along the Gulf Coast east and west of Louisiana and was sending expeditions north from Santa Fe into Colorado to counter French claims to the extent of Louisiana. That left England with a narrow strip of land between the Atlantic coast and the Appalachian Mountains.

All this time, the wars of Europe had continued to rage. Their causes were many: covetous territorial appetites, intense religious fervor, uncertain royal successions, and more and more frequently, commercial rivalries in an ever-expanding global marketplace.

Before the Treaty of Aix-la-Chapelle, Europe's wars had spilled over to North America, but they had been relatively minor sideshows. Three conflicts, however, became enough of an issue in North America that English colonists came to call them by the name of the reigning sovereign.

The War of the League of Augsburg (1688–1697) was called King William's War in the English colonies. It pitted an anti-French alliance that included England under William III, Sweden, Spain, Austria, Holland, and a few German states against Louis XIV, who had an appetite for Alsace and Lorraine. On the grander scale of European dynasties, it was the Hapsburgs versus the Bourbons. In North America, the French raided Schenectady, New York, and frontier settlements in New England—while forces from Massachusetts Bay captured Port Royal in Nova Scotia and brought the French governor back to Boston as a prisoner. Later, the English made a foray against Quebec but were beaten back in disarray. The Peace of Rijswijk (Ryswick) in 1697 restored all territory to its original claimants, but John Pynchon of Springfield, Massachusetts, spoke what remained on the minds of many New Englanders: "We shall never be at rest until we have Canada."[2]

Queen Anne's War (1702–1713) came about somewhat differently, but the protagonists were largely the same. It was called the War of the Spanish Succession in Europe, and the issue was just that. Who would sit on the Spanish throne when Charles II died without issue? Conveniently for Louis XIV of France, whose desire for all of Europe had not diminished, Charles II willed his Spanish throne to Louis's grandson, Philip of Anjou. France and Spain united! That set English hearts—and many more on the continent—aflutter. To counter the threat of this combined power, William III formed the Grand Alliance of England, Holland, Prussia, Austria, and most of the Holy Roman Empire states. When William III died in 1702, Queen Anne inherited both the English throne and the war.

In North America, because Spain was a nominal ally of France, there was conflict on both the northern and the southern borders of the English colonies. France's Abenaki allies burned Deerfield, Massachusetts, provoking resentment that would still be remembered two generations and two wars later. New Englanders once again captured Port Royal in Nova Scotia, but yet another English expedition against Quebec turned around and sailed back down the Saint Lawrence while still 100 miles short of its goal. South Carolinians burned Spanish Saint Augustine—but could not capture its presidio—and Indian nations throughout the south resisted incursions from both sides with ever-changing alliances. One grand French scheme to move north from the West Indies and "chase our adversaries from Carolina . . . insult New York, attack Virginia, [and] carry help to L'Acadie and Newfoundland" was stillborn when its leader and hundreds of soldiers died of yellow fever in Cuba.[3]

The war dragged on in Europe for so long that yet another royal succession became problematic. Emperor Joseph I of Austria died without issue and was succeeded by his brother, Charles, who up until then had been the candidate of the Grand Alliance to become king of Spain. Europe looked around and decided that the only thing worse for the global balance of power than France and Spain united under the Bourbons would be Spain and Austria (and assorted states of the Holy Roman Empire) united under the Hapsburgs. In the end, more warfare and Louis XIV's diplomacy finally achieved recognition for his grandson as Philip V of Spain on the condition that Spain and France never be united.

The resulting Treaty of Utrecht of 1713 had many provisions, but the chief result was to keep any one empire from dominating. In North America this meant that France yielded toeholds to England in Nova Scotia, in Newfoundland, and on the southern shores of Hudson Bay. England, which had used the war to build up its naval power while Louis XIV concentrated on his armies, also received Gibraltar and Minorca from Spain, as well as the right to participate in the slave trade in the New World.

"Before Queen Anne's reign," according to the naval historian Samuel Eliot Morison, "England was *a* sea power; after 1713 she was *the* sea power, and long so remained."[4] With the benefit of historical hindsight that is certainly true, but at the time England—or Great Britain, as the unified England, Wales, and Scotland were technically called after 1707—still had a war or two to fight before it would be the unchallenged ruler of the seas.* And nowhere was England more challenged than in North America. Historians have been quick to describe the quarter century after the Treaty of Utrecht as an era of peace, but that generalization hardly begins to account for the friction that continued along the English, French, and Spanish frontiers in North America and the surrounding seas.

In the north, the Treaty of Utrecht permitted France to retain the fortress of Louisbourg and Cape Breton Island. The French set about fortifying the island and keeping the loyalty of the Acadians on Nova Scotia, who were now nominally subjects of the new king of England, George I of Hanover. The French also built Fort Saint Frédéric on Lake Champlain south of Montreal. Crown Point, as the English would call it, was an annoying finger of French presence stuck down England's throat on the New York frontier. Farther west, the French built Fort Niagara on Lake Ontario at the mouth of the Niagara River below the falls and augmented their positions on the Mississippi with two more outposts on the Wabash River and the lower Ohio River. Pierre de La Vérendrye even built French forts—if the crude stockades could be called forts—on Lake of the Woods and the future site of Winnipeg. Clearly, France was reaching to embrace a continent.

In the south, French Louisiana became firmly established with the founding of New Orleans in 1718. This fact and Spain's frontier

---

* During the French and Indian War, the British Union flag, long called the Union Jack, bore the thick horizontal red cross of Saint George representing England and the narrow diagonal white cross of Saint Andrew representing Scotland, on a blue field. The narrower diagonal red cross of Saint Patrick representing Ireland was added to the banner after the Act of Union of 1800 incorporated Ireland into Great Britain. The French banner, the fleur-de-lis, was white with gold lilies.

in Florida prompted England to give General James Oglethorpe a land grant to settle a buffer state between the Carolinas and Florida. Oglethorpe founded Savannah, Georgia, in 1733. In time, Georgia became a royal province, but not without more than a little opposition from Spain. Philip V urged "extirpation of the English from the new colony of Georgia which they have usurped" and "laying waste [to] South Carolina and her dependencies."[5]

Then things really got bizarre. Friction between the English South Sea Company and Spanish monopolies over trading in the Caribbean became intense when an English sea captain named Jenkins—a smuggler at best, a pirate at worst—told a tale of being captured and having the tip of one ear cut off by a Spanish cutlass. By the time Jenkins was displayed before the House of Commons, the War of Jenkins' Ear was in full swing. Oglethorpe advanced to Saint Augustine with "1,600 men on seven warships carrying forty dugouts for landing operations," but failed to capture it. The Spanish retaliated by landing troops on Saint Simons Island on the Georgia coast and advanced on the settlement of Frederica before Oglethorpe managed to halt them at the battle of Bloody Marsh on July 7, 1742.[6]

By now, all of Europe was lining up for another major conflict over that most heated of topics—royal succession. This time, the throne was Austria's, but it was not exactly vacant. Shortly after the Treaty of Utrecht, Charles VI of Austria, who had no son, went scurrying around the courts of Europe with a document called the Pragmatic Sanction, which purported to ensure the territorial integrity of his empire when his daughter, Maria Theresa, eventually ascended to the throne.

As it turned out, Maria Theresa could more than take care of herself, but not without a fight or two. Charles VI of Austria died in October 1740, a few months after Frederick William I of Prussia. Frederick William's son, another Frederick, had no intention of respecting his father's endorsement of the Pragmatic Sanction and promptly marched into Austria's rich province of Silesia. Prussia and France became unlikely allies when Louis XV thought that France might win Austrian spoils in the conflict. England sided with Austria.

The ensuing War of the Austrian Succession (1740–1748) was called King George's War in North America after George II, the second of the Hanovers to rule Great Britain. Frederick, whom history would call "the Great," seemed to fight only when it suited him, withdrawing from the conflict after seizing Silesia, then reentering it several years later. Great Britain's policy was to cajole and subsidize its allies—principally Prussia and the Low Countries—to keep the French occupied in Europe while the Royal Navy and the cream of the British army picked the plums of France's overseas empire.

In North America, the usual frontier raids occurred, but the big news was in the North Atlantic. The royal governor of Massachusetts, William Shirley, determined to capture the French fortress of Louisbourg in Nova Scotia with a ragtag collection of New England militia. William Pepperell, a merchant from Kittery, Maine, led the troops with assistance from a Royal Navy squadron. By some accounts, the expedition more closely resembled a fraternity camping trip than a disciplined military campaign, but the result could not be denied.

Landing on the rocky coast of Cape Breton Island several miles from Louisbourg in April 1745, Pepperell's forces created such havoc and confusion that the French commander surrendered the fortress six weeks later. The following year, France sent a fleet of almost 100 ships to retake Louisbourg and burn Boston in retaliation, but Atlantic storms and the dread of scurvy turned the fleet before it had fired a shot. Meanwhile, the feuding continued in Europe until, momentarily worn out, the belligerents gathered at Aix-la-Chapelle.

By the time the bells rang in Charlemagne's ancient cathedral, they heralded peace—but a peace that did not resolve the geopolitical realities of colliding frontiers. The Treaty of Aix-la-Chapelle put most of the pieces back on the board where they had been at the beginning and merely signaled a short recess in the underlying disputes. Frederick II was allowed to keep Silesia, but that only hardened Maria Theresa's determination to beat him the next time. Indeed, the four most important signatories to the treaty—Great Britain's George II, France's Louis XV, Austria's Maria Theresa, and Prussia's Frederick II—would all be around the next time.

France gave Madras in India back to the English, and the English gave Louisbourg back to the French, much to the disgust of most New Englanders. The trading rivalries that had cost Jenkins his ear were not addressed, although one West Indian merchant, William Beckford, summed up what appeared to remain the mercantile policies of many: "Our trade will improve by the total extinction of theirs." So the Treaty of Aix-la-Chapelle became the "peace without victory." In France, it even inspired the expression, *Bête comme la paix*, "as stupid as the peace."[7]

So there the world stood in 1748. The bells of Aix-la-Chapelle were not so much a call to peace, as a warning to prepare for another, inevitable war. Tangling alliances, commercial rivalries, royal successions, and a litany of treaties all made for a complex and volatile situation. But this global background has much to say about what was unfolding in North America and why the names of Rijswijk, Utrecht, and Aix-la-Chapelle should come to be inscribed on lead plates buried along the Ohio River the following year.

2

BEAUTIFUL OHIO

After the Treaty of Aix-la-Chapelle, the map of North America still showed the English colonies heavily encircled by a French empire stretching from the mouth of the Saint Lawrence to the Mississippi delta. A closer look at the map, however, revealed a startling chink in the French armor. Quite logically, France had established its outposts along the major rivers between the Great Lakes and New Orleans to afford the easiest avenues of transportation and communication. This line ran roughly between the western Great Lakes and the Mississippi River via either the Lake Michigan–Illinois River portage or the Lake Erie–Maumee River–Wabash River portage, the latter of which was scarcely a dozen miles long, just south of present-day Fort Wayne, Indiana.

Between this line of French forts and the crest of the Appalachians 300 to 400 miles to the east, however, was a tremendous expanse of territory drained principally by the Ohio River and its tributaries. It was not a no-man's-land. The English were beginning to overflow the Appalachians, which had once dammed their expansion and kept it in check. France might claim vast territory, but the English colonies had one thing that New France did not—the population to fill the land. The census of 1754 showed about 55,000 white inhabitants in Canada, plus perhaps another 25,000 in Acadia and Louisiana. By comparison, the English colonies boasted an estimated 1,160,000 white inhabitants, plus some 300,000 black slaves.

In large part this disparity in population was because the English colonies were already somewhat of a western European melting pot. France, on the other hand, kept a close check on immigration up the Saint Lawrence, rigidly controlling the numbers politically and restricting them religiously to French Catholics. Even Protestant Huguenots, banished from Catholic France by the hundreds of thousands, found their way to the middle and southern English colonies rather than New France.

As Francis Parkman summed it up: "France built its best colony on a principle of exclusion, and failed; England reversed the system, and succeeded." No wonder, then, that traders and trappers and even a settler or two from the English colonies were crossing the Appalachians in increasing numbers and making themselves quite at home along the valleys and tributaries of what the French called the Belle-Rivière, "beautiful river."[1]

The Iroquois called the river the Ohio, meaning "something big." If relations between the French and English were complex and the rivalries of Europe convoluted, the status of Native Americans along the upper Ohio River was even more so. The dominant power in the region was the Iroquois Confederacy, a union of five nations—the Mohawk, Seneca, Oneida, Onondaga, and Cayuga—that coordinated external relations with other tribes as well as with the French and the English. When these five were joined by the Tuscarora, the empire of the Six Nations stretched from the upper Hudson River westward to the Ohio and was an immense buffer between the French and English frontiers.

As the power of the Iroquois Confederacy grew, it exerted what in European terms might be called a feudal domination over other tribes, including the Mingo, Shawnee, and Delaware. These tribes—some historians have simply lumped them together as the "Ohio Indians"—had generally been pushed westward by European settlements.

Despite numerous pronouncements of neutrality, the Six Nations were constantly wooed by both the French and the English. Historically, the Mohawk along the New York frontier were more likely to trade with and be influenced by the English, while the Seneca along lakes Ontario and Erie were more likely to look north to the French. Neutrality aside, the Six Nations referred to its commercial and strategic relationship with the English as the "Covenant Chain" and maintained a similar relationship with the French.[2]

While generally agreeable to advantageous trading relations, the Iroquois Confederacy, like most Native Americans, came to resist European encroachments that had an air of permanency. Passing coureurs de bois (trappers) in canoes were one thing; fresh-cut log cabins and planted fields were quite another. Of course, the French, whose empire was based largely on a transitory fur trade anchored at a few key points, were of a similar mind. Thus as the dust of Aix-la-Chapelle settled, the French determined to do something about these English incursions into the Ohio Valley and strengthen their relations with the Ohio Indians before it was too late.

In the spring of 1749, the governor of New France, the marquis de la Galissonière—who had already pleaded in vain with Louis XV for more colonists—dispatched Captain Pierre-Joseph Céloron de Blainville to reassert French claims to the upper Ohio River country. At age fifty-six, Céloron was no stranger to the Great Lakes and the French frontier. He had commanded posts at Michilimackinac, Pontchartrain, and Niagara and had most recently been commandant at Fort Saint Frédéric on Lake Champlain. On June 15, Céloron departed Montreal with a Jesuit chaplain, Father Bonnecamps, fourteen officers, and some 200 soldiers and Indian allies. His entourage ascended the rapids of the Saint Lawrence in twenty-three birch-bark canoes and three weeks later reached Fort Niagara on Lake Ontario. After a week portaging around Niagara Falls, the flotilla paddled along choppy Lake Erie until just west of present-day Dunkirk, New York, where it portaged again, this time south to placid Lake Chautauqua.

The idyllic waters of Chautauqua were short-lived, however, and despite heavy rains that had pelted them during the portage, the Frenchmen found its outflow stream to be but a trickle. "In certain places, which were only too frequent," recorded Father Bonnecamps in his journal, "there were barely two or three inches of water . . . [and] we were reduced to the sad necessity of dragging our canoes over the stones—whose sharp edges, in spite of our care and precautions, took off large splinters." Finally, on July 29, 1749, Céloron and his company reached the deeper waters of the Allegheny.[3]

Here, at the confluence of Conewango Creek and the Allegheny (present-day Warren, Pennsylvania), Céloron undertook his first official act. With his troops duly assembled before him, the captain nailed a sheet of metal bearing the coat of arms of Louis XV to a tree. At its base, Céloron buried a lead plate with an inscription proclaiming that this act was done "for a monument of renewal of possession which we have taken of the said river Ohio, and of all those [rivers] which fall into it, and of all those territories on both sides as far as the source of the said rivers."[4] Noble words, but not much of a deterrent to trespassers—particularly when covered by a few shovels of dirt.

Céloron's birch-bark fleet floated down the Allegheny, and he buried a second lead plate below its confluence with French Creek. This was near the village of Venango, where a long-established English trader named John Fraser and the town's inhabitants took to the woods at the approach of French arms. Another Indian village, Attiqué (near present-day Kittanning, Pennsylvania), was similarly deserted.

When the French overtook six English traders—Céloron called them "soldiers" in his journal—at yet another abandoned Indian village, Céloron expressed surprise, feigned or not, to find them trespassing on French soil and admonished them to leave the country or face unpleasant consequences. Céloron sent a letter with them to Governor James Hamilton of Pennsylvania asserting that "our Governor-General [Galissonière] would be very sorry to have to resort to violent measures, but he has positive orders not to allow foreign merchants or traders in his government."[5]

On down the Allegheny Céloron's flotilla went. When a huge, slow-moving river joined the Allegheny from the south, Céloron floated past without comment. This was the Monongahela, the Allegheny's mighty twin; and the narrow point of land at their confluence was empty. If Céloron was of a mind to bury plates, what could have been a more perfect location than this? The site would not remain empty for long.

Perhaps Céloron's nonchalance at the true beginning of the Ohio was a result of his haste to reach the Indian village of Chiningué, which the English called Logstown (near present-day Economy, Pennsylvania). Here, Céloron spent three days in council with its inhabitants, including ten English traders. Again, he delivered Galissonière's message: this was French land; the English traders must leave; and the Ohio Indians must have nothing further to do with them. But Céloron was fighting an uphill battle. It was really a matter of simple economics. Given their proximity to Virginia, the English traders were able to sell goods to the Indians for about one-quarter of the French price.[6]

Paddling down the Ohio, Céloron and his men buried lead plates at the mouths of Wheeling Creek, the Muskingum River, and the Kanawha River. By the time the expedition reached a large Shawnee village at the mouth of the Scioto River, Céloron was feeling rather outnumbered. He warned the village against the English, but in front of the assembled Shawnee he stopped short of claiming that their land belonged to France.

Finally, on August 31, 1749, at the mouth of the Great Miami River just west of present-day Cincinnati, Céloron buried his sixth and final plate. The last line asserted that France's right to this ground had been "maintained therein by arms and by treaties, and especially by those of Rijswijk, Utrecht, and Aix-la-Chapelle." But that meant little here. It had become painfully obvious in the course of their journey that the Ohio country, as Father Bonnecamps shrewdly observed, was "so little known to the French and, unfortunately, too well known to the English."[7]

From the mouth of the Great Miami, Céloron and his expedition followed the river north to the Miami village of Pickawillany (present-day Piqua, Ohio). Its chief was known as La Demoiselle by the French but was nicknamed Old Briton by the English, because of his close friendship with English traders. It was another painful reminder that while the French were busy burying lead plates, the English were sending an increasing stream of commerce into the Ohio country.

Céloron and his men abandoned their canoes at Pickawillany and marched overland to the principal Miami village at the headwaters of the Maumee River (present-day Fort Wayne). Here, they procured new canoes and floated down the Maumee to Lake Erie, eventually completing the circle back to Montreal. "All I can say," Céloron concluded in his journal on his return, "is that the nations of these countries are very ill disposed toward the French, and devoted entirely to the English."[8]

And the English certainly had designs on the country. By a variety of overlapping royal land grants and other claims, Massachusetts, Connecticut, New York, Pennsylvania, and Virginia each coveted parts of the Ohio Valley. Virginia was the most aggressive. Even as Céloron was burying his plates, Virginia was granting 200,000 acres of land west of the Appalachians to the Ohio Land Company on the condition that 100 families settle on it within seven years. Here was another difference between English and French colonial policies. England—just as it had done since the ill-fated Roanoke adventure—was pushing its frontiers with commercial settlements, a force that in the end proved far more unstoppable than banners and lead plates.

The Ohio Company quickly hired a surveyor, Christopher Gist, to peruse the Ohio Valley for the best lands. Gist was the original Daniel Boone—pathfinder, hunter, trader, and land developer. The end result of Gist's subsequent work was that between 1750 and 1754 the English strengthened their presence in the upper Ohio Valley and even received permission from the Ohio Indians to fortify the trading post at Logstown. That greater inroads were not made was due in part to incessant bickering among New York, Pennsylvania, and Virginia over which colony had the right to negotiate with the Six Nations and claim these western lands.[9]

However jumbled and disjointed these English advances, they continued to arouse great alarm in the French. In 1753, the new French governor, the marquis de Duquesne, recalled Paul Marin de la Malgue from his post at Baie-des-Puants (present-day Green Bay, Wisconsin) and ordered him to build a new string of forts that would serve to tighten New France's frontier along the upper Ohio.

Forsaking Céloron's route to the Ohio via Lake Chautauqua and its steeper portage, Duquesne ordered Marin to construct a fort on Lake Erie at Presque Isle (present-day Erie, Pennsylvania) and another about a dozen miles south of there at the head of French Creek. The latter post, called Fort Le Boeuf, was hardly more than an enclosed square of crude pickets, but here Marin's troops built boats and then floated down French Creek. On August 28, 1753, they seized the trading post at Venango on the Allegheny River that Englishman John Fraser had quietly reoccupied following Céloron's visit four years before. This time, Fraser retreated down the Allegheny.

If the English were indignant at these French advances, the Iroquois were even more so. "We don't know what you Christians, English and French together, intend," the Iroquois told the English Indian agent William Johnson; "we are so hemmed in by both that we have hardly a hunting place left. In a little while, if we find a bear in a tree, there will immediately appear an owner of the land to challenge the property, and hinder us from killing it, which is our livelihood."

But it was neither indignation nor force of arms that halted the French advance down the Allegheny. Rather, a general malady of assorted fevers, scurvy, and exhaustion combined with the approach of winter. By fall, of the more than 2,000 men who had begun the campaign, only 800 were fit for duty. Marin sent most back to Montreal and went into winter quarters at Fort Le Boeuf, where he abruptly died on October 29, 1753.[10]

If the remaining French at Fort Le Boeuf thought they were settling in for a quiet winter, they were mistaken. Just after the early sunset of December 11, 1753, a gangling young man appeared out of the woods accompanied by a grizzled old-timer and half a dozen others leading packhorses. The older man was Christopher Gist, by now no stranger along the tributaries of the Ohio. The younger man was unknown to the French. He spoke no French but managed to convey that he had come from Virginia with an important message. His name, he said, was George Washington. It meant nothing to them.

It is difficult to get a mental picture of George Washington at this age and in this environment. He was two decades away from leadership in the Revolution and even further from being called the father of his country. Rather, he was a tall and solid youth of twenty-one, a man by colonial standards, to be sure, but as yet untried in just about everything. What schooling he had received was marginal. It had been augmented by practical experience in the field as a surveyor and service as a major in the Virginia militia— the latter influenced by his half-brother Lawrence's military exploits during the War of Jenkins' Ear. Lawrence Washington named his estate in Virginia after the English admiral he served under; and shortly after Lawrence died in 1752, twenty-year-old George became the master of Mount Vernon.

While Marin had been marching toward the forks of the Ohio, the lieutenant governor of Virginia, Robert Dinwiddie—by no small coincidence a shareholder in the Ohio Company—had been badgering the crown to build its own chain of forts in the region. In the fall of 1753, Dinwiddie received permission to do so at the colony's expense, provided he first proclaim to the French "our undoubted rights to such parts of the said river Ohio, as are within the limits of our province of Virginia, or any other." Should such diplomatic entreaties fail, Dinwiddie was authorized to "repel force by force."[11]

Despite the lateness of the season, Dinwiddie chose young Major Washington to deliver a message to the French as soon as possible. Leaving Williamsburg, Virginia, on October 31, 1753, Washington went via Fredericksburg and Alexandria to the Ohio Company's trading post at Wills Creek (present-day Cumberland, Maryland) on the upper reaches of the Potomac River. Here, Washington retained the services of the indomitable Gist as his guide. Heavy rain and an early snow hampered them as they plodded onward to John Fraser's new trading post at the mouth of Turtle Creek on the Monongahela about ten miles upstream from its confluence with the Allegheny.

The swollen waters of the Allegheny forced Washington and Gist to borrow a canoe from Fraser to effect the crossing; and while they waited at the confluence for its arrival, Washington "spent some time in viewing the rivers and the land in the fork." The site, the young major noted, was "extremely well situated for a fort, as it has the absolute command of both rivers. The land at the point is 20 or 25 feet above the common surface of the water, and a considerable bottom of flat, well-timbered land all around it, very convenient for building."[12]

Once across the Allegheny, Washington and Gist continued two miles down the north bank of the Ohio to the location that the Ohio Company intended for a fort. Washington was not impressed. "As I had taken a good deal of notice yesterday of the situation at the forks," he recorded, "my curiosity led me to examine this more particularly, and I think it greatly inferior, either for defense or advantages; especially the latter; for a fort at the forks would be equally well situated on [the] Ohio and have the entire command of [the] Monongahela, which runs up to our settlements."[13]

Next, there was a stop at Logstown for a council with the Seneca chief Tanaghrisson, called Half King by the English. In addition to asking the French to leave the Ohio, Washington was charged with winning the Indians' support for England's claims, or at the very least denying such support to the French. Washington heard reports of the French presence on the Allegheny and exchanged wampum, the part-mystical, part-monetary strings of beads that were the recognized medium of trade and diplomacy among many Native Americans. Tanaghrisson proved a gracious host, and it was all that Washington could do to take leave without offending him. Expressing concerns for the young officer's safety, Tanaghrisson agreed to accompany him to meet the French.

Arriving at Venango on December 4, "without anything remarkable happening but a continued series of bad weather," Washington found the French flag flying from Fraser's old house. Captain Philippe de Joncaire politely said that while "he had the command of the Ohio," Washington's missive was best directed to a general officer at Fort Le Boeuf. Joncaire, too, proved a gracious host and Washington observed that over the course of dinner, "the wine, as they dosed themselves pretty plentifully with it, soon banished the restraint which at first appeared in their conversation, and gave a license to their tongues to reveal their sentiments more freely. They told me, that it was their absolute design to take possession of the Ohio, and, by G—, they would do it."[14]

The following morning, heavy rain prevented Washington's departure for Fort Le Boeuf, but Joncaire took advantage of it by wining and dining Tanaghrisson, a situation that made Washington quite nervous. He was supposed to be winning Indian allies, not affording the French an occasion to do so. When on the morning of December 7, Washington was able at last to head north for Fort Le Boeuf, the journey took four days. It was slowed, reported Washington, by "excessive rains, snows, and bad traveling, through many mires and swamps, which we were obliged to pass, to avoid crossing the creek, which was impossible, either by fording or rafting."[15]

Washington arrived at Fort Le Boeuf after dusk on December 11 and was conducted the next morning into the presence of Marin's newly arrived replacement, Captain Jacques Legardeur de Saint Pierre. The fifty-two-year-old captain had been at his post for only a week. Washington described him as an "elderly gentle-

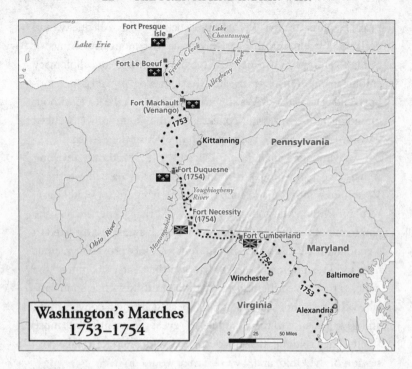

Washington's Marches 1753–1754

man" with "much the air of a soldier." After waiting for the arrival of another officer from Fort Presque Isle, Legardeur accepted Dinwiddie's letter and adjourned into a private room to study it.

Dinwiddie's letter charged that "the lands upon the River Ohio, in the western parts of the Colony of Virginia, are so notoriously known to be the property of the Crown of Great-Britain, that it is a matter of equal concern and surprise to me, to hear that a body of French forces are erecting fortresses, and making settlements upon that river, within his Majesty's dominions." The governor was sending Washington, the letter continued, "to complain to you of the encroachments" and "require your peaceable departure." The latter was, of course, wishful thinking at best, naïveté at worst. But in the manner of the times, Dinwiddie could not help hoping that the French would "receive and entertain Major Washington with the candor and politeness natural to your nation" and "return him with an answer suitable to my wishes for a very long and lasting peace between us."[16]

Legardeur took a day to consult with his officers before crafting a diplomatic but equally firm rebuttal. Noting that he was but a subordinate and wishing that Washington's orders had required him to "proceed to Canada, to see our General [Duquesne]," Legardeur nonetheless assured Dinwiddie that he would transmit the letter to Duquesne. Meanwhile, of course, the Frenchman was hardly inclined to honor Dinwiddie's request that he withdraw. "I do not think myself obliged to obey it;" Legardeur replied in a letter to Dinwiddie, and "whatever may be your instructions, I am here by virtue of the orders of my general; and I entreat you, sir, not to doubt one moment, but that I am determined to conform myself to them."[17]

That didn't leave much for Washington to say. He hurriedly departed Fort Le Boeuf for Virginia before the vagaries of the weather and the hospitality of the French toward his Indian companions could delay him any longer. Having already sent the packers and their horses ahead, Washington and Gist floated down French Creek in a canoe to Venango, plagued by shoals and ice most of the way. From there, they struck overland for the forks, expecting to find the Allegheny frozen solid. It wasn't. After a full day spent building a raft with "but one poor hatchet," Washington and Gist pushed off from shore only to be immediately caught in a morass of floating ice.

Washington attempted to fend off some of the larger floes with a pole, but the current was moving so swiftly that it snatched the staff and jerked Washington into the frigid water. He managed to save himself by grabbing one of the protruding logs of the raft, but despite all their subsequent efforts, the two men could not steer the improvised vessel to the opposite shore. As the river swept by an island, Washington and Gist jumped for it and splashed to soggy ground. A bitter and most unpleasant night followed, but in the morning the river was frozen enough to permit them to reach the opposite shore. By noon they were safe at

Fraser's Monongahela trading post. After that sort of adventure, the remainder of their journey was decidedly anticlimactic.

On his return to Williamsburg, Washington hurriedly wrote a report for Governor Dinwiddie from notes that he had kept. He had but one day to do so, because Dinwiddie immediately summoned the general assembly to hear the latest news from the Ohio and ordered Washington's report printed for distribution. Apologizing for "numberless imperfections" in what he thought was only for Dinwiddie's perusal, Washington expressed surprise that his journal "would ever be published, or even have more than a cursory reading."[18]

Little did Washington know at the time that his own role in another conflict would give his account a huge audience and that subsequent events would mark his journey to the waters of the beautiful Ohio as the beginning of the ultimate contest for a continent.

3

ALBANY, 1754

As Virginia pondered the ramifications of George Washington's initial reception on the Ohio, Great Britain's northern colonies were also growing increasingly uneasy with their French neighbors. Nowhere was this more a topic of conversation than in the frontier metropolis of Albany, New York. For almost a century, Albany had been the Gibraltar of the English colonial frontier. The explorer Henry Hudson first proposed its location just south of the confluence of the Hudson and Mohawk rivers in 1609. Begun by the Dutch as Fort Orange in 1624, the outpost and subsequent settlement of Beverwyck passed to the English along with New Amsterdam in 1664. Albany quickly became the linchpin that anchored New England's frontier running eastward to the Atlantic and the fulcrum that leveraged New York's expansion westward toward the Great Lakes.

In the other directions, Albany sat squarely on the north-south water corridor between New York City and Quebec via Lake Champlain and the Hudson and Saint Lawrence rivers. During all the French and Indian conflicts of the seventeenth and eighteenth centuries, Albany was frequently the only secure position on the English colonies' northern frontier. Later, during the American Revolution, Albany was still a pivot point, and three British armies—one of which was compelled to surrender at Saratoga—sought to converge on it.

In addition to its strategic military position, Albany in the spring of 1754 was well established as the commercial hub of the upper Hudson River valley and the center of the English fur trade. The small, wooden structure of Fort Orange had been abandoned long ago to the annual floods of the Hudson, and the English had built a more substantial fort of thick stone walls on higher ground above the western end of State Street.

English rule, however, had done little to change Dutch custom or lineage. Albany's houses were "very neat, and partly built of stones covered with shingles of white pine." Most inhabitants were of Dutch descent, still spoke Dutch, listened to sermons preached in Dutch, and possessed impeccable Dutch manners. "In the evening the verandas are full of people of both sexes;" observed one traveler, "but this is rather troublesome because a gentleman has to keep his hat in constant motion . . . [as] it is considered very impolite, not to lift your hat and greet everyone." Recorded populations of 714 in 1697 and 1,128 in 1714 had grown to "more than three thousand inhabitants" by 1754.

Two imposing churches occupied opposite ends of a wide marketplace through which State Street ran between the fort and the Hudson River. The Dutch Reformed Church sat near the river at the eastern end of the marketplace and was built of stone capped by a small steeple with a bell. "The English church," Saint Peter's Episcopal, sat directly below the fort at the western end of the marketplace. It, too, was built of stone, but it lacked a steeple.

The other grand structure in Albany was the city hall, located near the river just south of the Dutch church. The hall was "a fine building of stone, three stories high" with a steep-pitched roof and "a small tower or steeple, with a bell, and a gilt ball and vane at the top of it." The bell dutifully tolled the hours of noon and 8:00 p.m. each day. After the fort and the Dutch church, the city hall was the town's largest structure, plainly visible on the Albany skyline to arriving visitors.[1]

Now, in June 1754, there were important visitors arriving in town. Albany had long been the scene of conclaves between the

English and representatives of the Six Nations, and with the French once again striving to win the latter's allegiance, it was high time for another such gathering. The previous September, Great Britain's Board of Trade—essentially the governing body for his majesty's colonies and by virtue of its very name an indication of how closely commercial and imperial interests were tied—had announced an appropriation for "presents" for the Six Nations. It also directed the governor of New York to convene a conference with their representatives "for delivering those presents, for burying the hatchet, and for renewing the Covenant Chain."[2]

The concept of "presents" was a cornerstone of eighteenth-century diplomacy between all Native American nations and both the French and the English. These outright gifts included "fabrics, hardware, munitions, food, toys, jewelry, clothing, wampum, and liquors." Regarded most cynically, they were outright bribes. Viewed differently, "presents" were the most aggressive marketing inducements of the age, designed to win and keep commercial and political relationships.

Both France and England expended large sums to provide these gifts; but just as English goods were cheaper in trade, English presents were more abundant and of higher quality because of the proximity of the commercial hubs of New York, Pennsylvania, and Virginia. In the struggle for a continent, the Native American tradition of giving and receiving gifts as part of any negotiation had been taken to its extreme and "presents" were essential to any relationship.[3]

But there was also another issue on the table for delegates arriving in Albany that summer. The earl of Holderness, the British secretary of state charged with North America, had written to colonial governors late in the summer of 1753, urging them to be "in a condition to resist any hostile attempts that may be made upon any parts of His Majesty's dominions within your government." It was also the king's directive that "his provinces in America should be aiding and assisting each other, in case of any invasion."[4]

Governor William Shirley of Massachusetts, who had already

proved adept in colonial military matters by orchestrating the capture of Louisbourg during King George's War, enthusiastically took up this charge. Shirley assured his fellow governors that "in case any hostile attempts shall be committed upon His Majesty's territories," Massachusetts would respond. But Shirley went much farther. Nothing would be more effective in countering French movements and bringing the Indians to dependence on the English, Shirley told the Board of Trade in January 1754, "than a well concerted scheme for uniting all His Majesty's colonies upon it in a mutual defense of each other."[5]

By the time Governor Shirley addressed the Massachusetts general assembly on April 2, 1754, and asked it to appoint commissioners to the conference in Albany, he was even more emphatic. Citing frequent French attempts to "draw off the Six Nations from the English interest into their own," Shirley argued that nothing would be a more effective counter in maintaining Indian alliances than "one general league of friendship, comprising all His Majesty's colonies." Such a coalition of colonies might also "lay a foundation for a general one among all His Majesty's colonies for the mutual support and defense against the present dangerous enterprises of the French on every side of them."

Indeed, this was not just a problem for Massachusetts, Shirley continued: "Close on the back of the settlements of His Majesty's southern colonies, they [the French] are joining Canada to the Mississippi by a line of forts and settlements along the great lakes and rivers, and cutting off all commerce and intercourse between the English and the numerous powerful tribes of Indians inhabiting that country," all the while stirring up their resentment.[6]

There was, of course, another side to the matter. The French were equally convinced that it was the English who were plotting to seize the continent solely for their own purposes. As long ago as Céloron's visit, the French commandant at Fort Miami had bitterly noted that "the English spare nothing to keep [the Indians] and to draw away the remainder of those who are here. The excessive price of French goods in this post, the great bargains which the English give, as well as the large presents which they make to the tribes, have entirely disposed those tribes in their favor. . . . We have made peace with the English, but in this country they do not cease working to make war on us by means of the Indians."[7]

But whether against the French, the Indians, or both, some Englishmen clearly thought that some form of confederation was at hand. "For forming this general Union, gentlemen," Shirley told the Massachusetts assembly in his concluding remarks, "there is no time to be lost. The French seem to have advanced further toward making themselves masters of this continent within these last five or six years, than they have done ever since the first beginning of their settlements upon it."[8]

An English colonial confederation—particularly in the cause of common defense—was hardly a new idea. As early as 1643, the fledgling colonies of Massachusetts Bay, Plymouth, Connecticut, and New Haven formed the New England Confederation to settle boundary disputes and provide for mutual defense. (Subsequent charters merged New Haven with Connecticut in 1665 and Plymouth with Massachusetts in 1691.) Each colony appointed two commissioners who met annually and even held the power to declare war, provided that three out of four colonies agreed.

The New England Confederation avoided war with the Dutch in the Hudson Valley by negotiating a treaty with Peter Stuyvesant in 1650, but almost fell apart three years later when Massachusetts Bay resolutely refused to join a subsequent military action against the valley. In 1675, the confederation unanimously declared war against the Wampanoag in King Philip's War. The formal New England Confederation ended a few years later—in part because of Massachusetts Bay's emerging dominance—but that did not stop informal cooperation. In April 1690, a conference in New York attracted representatives from New York, Massachusetts Bay, Plymouth, and Connecticut to rally troops for King William's War. Even Maryland sent a letter offering 100 men, while Rhode Island pledged money in lieu of troops.

During Queen Anne's War, delegates of Massachusetts, New Hampshire, Connecticut, and Rhode Island met to bemoan the crown's lack of support for a plan to attack Montreal with a colonial force that had assembled at Albany. New York was too frustrated by the delay to participate, but the delegates drafted a petition to Queen Anne urging that a similar effort be mounted the following year. By the time of King George's War, Governor George Clinton of New York routinely summoned Iroquois chiefs to meet with colonial agents at Albany to distribute presents and maintain the Covenant Chain. Aware of those ritual gatherings, a businessman in Pennsylvania named Benjamin Franklin mused in 1751 that "were there a general council formed by all the colonies, . . . everything relating to Indian affairs and the defense of the colonies, might be properly put under their management."[9]

Born in Boston in 1706, Benjamin Franklin was the tenth son of a soap maker. He learned early that if greater opportunity was ever to knock, he had best be at the door waiting. Accordingly, he left Boston for Philadelphia at the age of seventeen and was soon apprenticed as a printer. Five years later, he was the proprietor of the *Pennsylvania Gazette*. By 1734, not only had the *Gazette* become "extremely profitable," but Franklin had secured the printing contracts for the colonial governments of Pennsylvania, Delaware, and New Jersey and launched *Poor Richard's Almanack*, a perennial best-seller.

This business success gave Franklin time to promote community causes that came to include Philadelphia's first subscription library, its first fire company, the Pennsylvania Hospital, and the University of Pennsylvania. Appointed his majesty's deputy postmaster general for North America in 1753, Franklin reorganized the colonial postal system and in doing so perhaps did more than any individual to tie together the distances and diversities of the English colonies.[10]

In May 1754, Franklin received word that gave increased urgency to his previous musings about mutual defense. Lieutenant

Governor Robert Dinwiddie of Virginia had dispatched Major George Washington back to the forks of the Ohio with about 150 men to support an expedition of the Ohio Company charged with finally constructing an English stronghold at the confluence of the Allegheny and Monongahela rivers. It had been too late. A much larger French force had already descended the Allegheny from Venango, captured the initial workings, and sent the Ohio Company's men packing.

Franklin reported this confrontation on May 9, 1754, in the *Pennsylvania Gazette* and used the occasion to publish on the same page what has frequently been called America's first political cartoon. The woodcut drawing showed a writhing snake severed into pieces labeled with the names of the individual English colonies. The admonishment in the caption was "Join, or Die."[11]

Franklin's cartoon was quickly republished in other papers, and many of them voiced similar concerns. "The making of an establishment on the River Ohio, is no new or partial scheme of the French, merely for the sake of trade, or a settlement on the lands," editorialized a New York newspaper, "but a thing long ago concerted, and but part of a grand plan for rendering themselves masters of North America."[12] And a newspaper as far away as Charleston, South Carolina, asserted that it was "greatly to be wished that the British provinces would unite in some system or scheme for the public peace and safety. Such an union would render us respected by the French, for they are not strangers to our power . . . and they will hardly make any experiment of our strength."[13]

So the distinguished visitors arrived in Albany and convened at city hall on June 19, 1754. There were twenty-three men sent by seven colonies: Massachusetts, New Hampshire, Connecticut, Rhode Island, Pennsylvania, Maryland, and New York. Two others—Virginia and New Jersey—were invited but did not send delegates, the former judging itself too preoccupied with events on its western border and the latter too insulated from them. Rhode Island and Connecticut were not invited—perhaps because of their known aloofness from matters of the interior—but sent delegates anyway, apparently at the urging of William Shirley. North Carolina, South Carolina, and Georgia (Georgia was then only barely under way as a royal colony) were not invited, notwithstanding some strong sentiments on the subjects to be discussed.

Among the delegates were Stephen Hopkins of Rhode Island, later to be its governor and a signer of the Declaration of Independence; Meshech Weare of New Hampshire, speaker of its assembly and later governor; Oliver Partridge of Massachusetts, later a member of the Stamp Act Congress; William Johnson of New York, already the leading expert in the colonies on Indian affairs and soon to play an even larger role; Thomas Hutchinson of Massachusetts, longtime speaker of its general assembly; and, of course, Benjamin Franklin of Pennsylvania. Colonial America would not see the like of such an assemblage again until the First Continental Congress convened some twenty years later.[14]

The negotiations at Albany with the Six Nations, whose representatives arrived shortly, proved to be typically labored and rife with ritual gifts. Finally, speaking for the Iroquois, Chief Hendrick asked for the reappointment of William Johnson as Indian commissioner to facilitate a clearer chain of communication and then turned to the matter of encroachments on Indian lands. It wasn't just the French who were encroaching, Hendrick complained, but men from Pennsylvania and Virginia as well. What was worse, the chief said, non-Iroquois Indians from Canada came to trade at Albany and other frontier posts and then left with "powder, lead, and guns" that circled back so that "the French now make use of [them] at the Ohio." Whose side were the English on? Hendrick demanded.

The delegates proceeded to assure the Iroquois that despite whatever recent neglect they might feel, the English valued the Covenant Chain and wanted the Iroquois tightly bound to his majesty's realm. In retrospect, what happened next appears somewhat incongruous with those assertions, although it was intended to appease the Indians regarding at least some of the encroachments. For a trifling 400 pounds, delegates from Pennsylvania purchased from the Iroquois a huge tract of land extending westward from the Susquehanna River in a great V.

A few days later, delegates from Connecticut working surreptitiously made their own deal for Connecticut's Susquehanna Company. For 2,000 pounds they purchased 5 million acres from the Iroquois between the Susquehanna's upper reaches and its western branch—some of which were also claimed by Pennsylvania. The delegates from Pennsylvania were furious; the delegates from Connecticut were smug. Only the Iroquois were momentarily satisfied, and they left Albany loaded down with thirty wagons of presents. Needless to say, these land transactions did nothing to promote a spirit of cooperation among the colonies.[15]

Meanwhile, during the long breaks in negotiations with the Six Nations, the other topic of the assemblage was roundly debated. Back in 1751, Benjamin Franklin had commented on the effectiveness of the Iroquois Confederacy to a friend. "It would be a very strange thing," Franklin had then remarked, "if six nations of ignorant savages could be capable of forming a scheme of such an union . . . and yet that a like union should be impracticable for ten or a dozen English colonies, to whom it is more necessary."[16] Now, it was up to the colonies to decide if that was true. By no small coincidence, Franklin arrived in Albany with notes in his pocket entitled "Short Hints towards a Scheme for Uniting the Northern Colonies."

Franklin's ideas, however, were not the only ones. His colleague from Pennsylvania, Richard Peters, proposed grouping the colonies into four geographic divisions that would hold annual meetings of various committees. Thomas Hutchinson of Massachusetts, with whom Franklin seems to have had detailed discussions on the subject, suggested forming the colonies into two unions, one north and one south, with the seat of government in the north being in Massachusetts. Thomas Pownall, who was not a delegate but was an ad hoc member of the Pennsylvania delegation, had its ear enough to advocate a strong anti-French plan that called for the construction of fleets and forts to gain control of lakes Erie, Ontario, and Champlain. But it was Franklin's ideas that bubbled to the top. "For though I projected the plan and drew it," Franklin later confided, "I was obliged to alter some things contrary to my judgment, or should never have been able to carry it through."[17]

When the delegates finally voted to "form a plan of Union of the Colonies by Act of Parliament," Franklin's fundamental concepts had survived, as well as his conviction that only an act of Parliament could meld the divergent interests and compel such a union. In hindsight, this vote looms as particularly significant. It specifically called for "an Act of Parliament." Clearly, this was no rogue assemblage rebelling against the British crown, but rather loyal subjects convinced that only Parliament "possessed the authority to alter the basic constitutional arrangements within the Empire"—something that would be sweepingly denied by many of the same men two decades later. The men at Albany were not seeking independence from England—far from it. Rather they were seeking royal protection from the French and some measure of assurance that the crown would pay for it.[18]

Franklin was slightly more practical and less constitutional in advocating action by Parliament. "Everybody cries, a union is absolutely necessary," Franklin wrote, "but when they come to the manner and form of the Union, their weak noodles are perfectly distracted. So if ever there be an Union, it must be formed at home [in England] by the ministry and Parliament."[19]

The delegates at Albany adjourned on July 11, 1754. They carried home to their respective colonial assemblies for ratification copies of what came to be called the Albany Plan of Union. In many respects, it was a most prophetic and prescient document. No one—save perhaps the shrewd Franklin—knew it, but in 1,169 words debated for just three weeks, the delegates had crafted the broad framework for what would become the Constitution of the United States of America about thirty years later. At the time, however, if the Albany Plan had been acted on by Parliament, it might just as easily have become the model for British colonial government.

Fully recognizing the sovereignty of King George II—and indeed desiring to maintain it—the delegates sought an act of Parliament "by virtue of which one general government may be formed in America, including all the said colonies." This general government was to be administered by a president-general appointed and funded by the crown. The president-general's legislative counterpart was to be a grand council elected by the assemblies of the individual colonies, not in proportion to population but rather in proportion to tax revenues actually paid to the general government. The result was similar to population-based representation in that it was weighted in favor of the more populous colonies—more people meant more taxes—but such an apportionment system was seen as an incentive for all but the smallest colonies to pay their tax share.

Of a total of forty-eight representatives, Massachusetts and Virginia were initially assigned seven each, while New Hampshire and Rhode Island were given two each. Representatives were to be elected for three-year terms with reapportionment based on tax revenues calculated once each cycle. No colony could ever have more than seven or less than two representatives.*

The grand council was to meet the first time in Philadelphia at the call of the president-general and then convene annually or more frequently if required by the representatives or summoned in emergencies by the president-general. The grand council was empowered to choose its own speaker, who was to serve as the president-general in the event of a vacancy in that office until the crown appointed a replacement. Members of the grand council

---

* The total apportionment was Massachusetts, 7; New Hampshire, 2; Connecticut, 5; Rhode Island, 2; New York, 4; New Jersey, 3; Pennsylvania, 6; Maryland, 4; Virginia, 7; North Carolina, 4; and South Carolina, 4. (Georgia was just becoming an official crown colony and Delaware had its own assembly but shared Pennsylvania's governor.)

were to be paid ten shillings per diem while in session or traveling to and from the council, with twenty miles judged to be a day's travel.

All legislative acts of the grand council required the consent of the president-general, who was charged with carrying them out. The president-general alone was to have the power to declare war on Indian nations and make treaties with them, the latter subject to the advice, but not the consent, of the grand council. For its part, the grand council was given the power to regulate Indian trade, purchase Indian lands beyond the current colonial boundaries, grant new settlements on such purchases, and govern those new settlements until "the crown shall think fit to form them into particular governments."

The grand council was also to have the power to "raise and pay soldiers and build forts for the defense of any of the colonies, and equip vessels of force to guard the coasts and protect the trade on the ocean, lakes, or great rivers." In doing so, however, the delegates recognized a strident political issue, over which a fledgling United States would fight a second war for independence half a century hence. They expressly prohibited the impressment of men into military service in any colony without the consent of that colony's legislature.

The grand council was given the power to "lay and levy such general duties, imposts, or taxes, as to them shall appear most equal" and—here one definitely feels Franklin's presence—collect such taxes "with the least inconvenience to the people; rather discouraging luxury, than loading industry with unnecessary burdens." No appropriations could be made from these funds, however, except by the joint orders of the president-general and the grand council.

The delegates further resolved that laws enacted by the grand council should be "as near as may be, agreeable to the laws of England." These were to be transmitted to the king in council for approval and be deemed accepted if not disapproved within three years after presentation. Perhaps most important to the looming crisis with France, the Albany Plan assured the individual colonies that their existing military and civil establishments remained intact and that in the event of attack, any colony might defend itself and have the expenses of such defense reimbursed by the general government.[20]

Charged with renewing the Covenant Chain with the Iroquois, the delegates to Albany had done that and much, much more; but how Parliament and their individual state assemblies would view the results of their labors was an entirely different matter. New Jersey and Connecticut were immediately jealous because the plan diluted their individual authority, particularly by the crown's appointment of a president-general. Rhode Island, despite Stephen Hopkins's assertion that such a union was an "absolute necessity," was also opposed. Maryland's governor, Horatio Sharpe, expressed displeasure that the plan of union had been forwarded to the Board of Trade before any colonial assemblies had had an opportunity to debate it. Maryland's general assembly consequently chose to ignore it, as did New Hampshire's and South Carolina's. North Carolina's governor promoted the plan largely as a revenue measure, but his assembly, too, was not persuaded.

That left the big four: Massachusetts, Pennsylvania, Virginia, and New York. After Franklin, Governor Shirley of Massachusetts was perhaps the most vocal advocate of the union—in part because he truly believed in it, but also because he may have seen himself as the logical crown appointee to be president-general. Still, Shirley was forced to confess to newly appointed governor of Pennsylvania, Robert Hunter Morris, that "I have no leaf in my book for managing a Quaker Assembly."

The best advice that Shirley could give Morris was to "lose no time for promoting the Plan of Union . . . as soon as possible." But in the Pennsylvania assembly with its majority of Quakers, it was hard to be in favor of common defense, when one was religiously opposed to the force that made that defense necessary in the first place. Pennsylvania waited until Franklin was absent from a session and quietly buried the plan.

In Virginia, where Governor Dinwiddie and the assembly had been too preoccupied with Virginia's western frontier to send representatives to Albany, neither governor nor assembly was in any rush to endorse a plan that promoted centralized control of the western lands that they were trying their utmost to grab. New York seemed inclined to support the plan, but never voted on it. Only in Massachusetts, the one colony that had given—at Governor Shirley's strong urging—full powers to its delegates at Albany to enter such a union, was the issue vigorously debated. Despite Governor Shirley's pleas to the contrary, the Massachusetts general assembly finally rejected the measure in January 1755.

Parliament, which might have avoided such colonial reactions by itself enacting some form of mutual defense measure—no matter how watered down—chose not to address the entire issue. As usual, it was Ben Franklin who cut to the chase and summed up the entire situation with just one sentence. "The assemblies did not adopt it," Franklin wrote some years later of the Albany Plan, "as they all thought there was too much [royal] prerogative in it; and in England it was judged too much of the democratic."[21]

Just how far apart the colonies were among themselves—even without the issue of a chief executive appointed by the crown—is evidenced by the fact that when they finally approved a confederation a generation later, it granted the central government much weaker powers than the plan championed by Franklin at Albany.

First submitted to the states by the Continental Congress in 1777, the Articles of Confederation held each state resolutely sovereign except for certain limited powers expressly delegated to Congress. Significantly, these did not include the power of taxation. But even then, there was not to be any quick ratification. New Jersey and Delaware withheld ratification until 1779 over objections to the vast western land claims of other states. Particularly jealous of Virginia's land claims, Maryland held out two years longer and ratified the Articles of Confederation only in 1781 after all the states ceded their western lands to the central government. That government evolved into its current federal form when the Constitution was finally ratified seven years later.

There is an often-told story of Benjamin Franklin, by then the undisputed elder statesman of the age, rising before the closing session of the Constitutional Convention in 1787 and pointing to the sun that was carved on the chair of the presiding officer, George Washington. He had often wondered during the rancor of the preceding debates, Franklin remarked, whether it was a rising or setting sun. Now, he was certain, he said, that the orb was rising.

But if one looks at the scene more closely and contemplates the many trials and tribulations that led from Albany in 1754 to Philadelphia in 1787, one can almost hear Franklin chortling to himself: "Apportioned representation, separation of powers, legislative advice on treaties. What good ideas! I'm glad I thought of them once before!" But for the present, in Albany in the summer of 1754, that day was far, far away, and the immediate road ahead lay filled with tremendous uncertainty.

4

BRADDOCK'S ROADS

The roads from Albany in the summer of 1754 led many places, but by the following spring the principal ones all belonged to the British major general Edward Braddock. "Few generals perhaps," a contemporary historian would soon opine, "have been so severely censured for any defeat, as General Braddock for this."[1] While occasionally referred to as the battle of the Monongahela, "this" has been far more frequently called "Braddock's defeat." Indeed, not until the golden-haired Custer failed to emerge from the Little Bighorn more than a century later would another leader's defeat be so personalized.

But the rough-cut road that led Edward Braddock to the Monongahela in the summer of 1755 was just one of four in a grand scheme that he was charged with overseeing to secure the expanding borders of King George's North American colonies. History has long debated whether he was up to the task.

In part, this debate has persisted because there is relatively little to say about Edward Braddock until he disembarked at Hampton Roads, Virginia, on February 20, 1755, and got his first look at North America. He was short in stature, of ample girth, and, at age sixty, by any standard of the time an old man who should have been content to sit by the fireside and glory in tales of his regiment. He was in fact the son of another Major General Edward Braddock, who had commanded the fabled regiment of the Coldstream Guards in Flanders and Spain during the War of the Spanish Succession. Young Edward joined his father's regiment in 1710 and slowly but surely worked his way up through the ranks, earning his promotions by longevity and loyal service rather than by purchase as was frequently the custom.

By 1745, the younger Braddock was lieutenant colonel of the Coldstream Guards, and he led the regiment in action on the continent the following year. It was in this service that he gained favor with the duke of Cumberland, who was George II's younger son and the commander of the British army. Much better known for zealous military action than diplomacy, Cumberland had just returned from brutally smashing the last gasp of the Stuart dynasty at Culloden Moor in the Scottish highlands. Braddock was subsequently named colonel of the Fourteenth Foot Regiment, and he joined it at Gibraltar in 1753. Within a year he was promoted to major general and as such was readily available a few months later when Cumberland wanted to impose a military solution on the disarray of Great Britain's North American frontier.[2]

In the summer of 1754 that disarray was being felt as abject panic in the offices of Virginia's lieutenant governor Robert Dinwiddie, in Williamsburg. News had come from the Ohio country that George Washington's second military mission to the forks of the Ohio had ended in humiliating defeat. The irony of the date would not become clear for almost a quarter of a century, but on July 4, 1754, Washington surrendered the crude stockade of Fort Necessity, about fifty miles south of the forks, to French forces and humbly retreated eastward to the Potomac. It had taken five years since Céloron's less than auspicious journey, but the French were now decidedly in control of the Ohio country, and even the Iroquois hastily sent emissaries to mend relations with them.

Dinwiddie reported this calamity to the Board of Trade and other high-ranking officials, but even before receiving Dinwiddie's official version, the de facto prime minister Thomas Pelham-Holles, the duke of Newcastle, warned that "all North America will be lost" unless the English countered French claims. Having said that, however, Newcastle favored conducting a neat and tidy, limited operation that would quickly and quietly secure key points on the English frontier before the French could send reinforcements and escalate matters into a full-scale war. That Newcastle leaned heavily on the less than tactful Cumberland to supply the military muscle for the operation was necessary, if somewhat ill-advised. Cumberland turned to Edward Braddock, perhaps—as some have suggested—because Braddock at sixty still desperately needed a job, but also no doubt because Braddock, too, was not one to allow tact to interfere with results.[3]

Whatever his strengths and weaknesses, General Braddock arrived in America with sweeping orders. Newcastle's neat and tidy approach had developed with Cumberland's involvement into a grand offensive scheme. Absent a formal declaration of war, Great Britain would nonetheless simultaneously attack the French frontier at four points. Given the communication and transportation networks of the time, "simultaneously" was, of course, a relative term.

Two regiments newly raised in the colonies and named after the heroes of Louisbourg in the previous war—Shirley and Pepperell—were to march west along the Mohawk River and seize Fort Niagara. Other colonials from New England, New York, and New Jersey were to strike north up the Hudson and rid the English of the annoying French finger of Fort Saint Frédéric at Crown Point. Still other New Englanders were to attack French outposts in Acadia and perhaps even seize the prize of Louisbourg that the English had given up once before. The fourth column, comprising two regiments of British regulars augmented by colonials and led by General Braddock himself, was to follow Washington's footsteps back to the forks of the Ohio and evict the French from the newly built Fort Duquesne.

As noted above, there was no declaration of war, and so the British claimed—apparently with straight faces—that these movements were simply designed to expel the French from lands that legitimately belonged to England. This had some degree of truth

based on the Treaty of Utrecht in the case of Acadia and even Fort
Duquesne, La Salle's grand claim of the drainage of the Mississippi
notwithstanding; but it didn't stand up very well with regard to the
other two locations. The French, after all, had occupied Fort Niagara for seventy years and Fort Saint Frédéric for twenty-four.[4]

Both sides, it seems, were not yet thinking globally. Each
naively thought that it could fight a limited war restricted geographically to North America. The duke of Newcastle—hardly
the shrewdest person ever to govern England—sensed a lack of
popular support for any war an ocean away when his countrymen
had had their fill of countless others just across the channel.
"Ignorant people," the duke noted—apparently not counting himself in that characterization—"say what is the Ohio to us, what
expense is there likely to be about it, shall we bring on a war for
the sake of a river."

The French were equally wary of an all-out war. As Braddock
arrived in Virginia, French commanders in the Ohio Valley received
word that "His Majesty [Louis XV] is on his side very far from
allowing that any invasion be undertaken against his neighbors" and
that any operations should be "strictly on the defensive."[5]

But it wasn't going to happen that way. Talk of limited war
while the troops engaged were mostly colonial militia was one
thing, but the distinction became quite muddy when both British
and French regulars were committed to the field. Indeed, the fic-
tion of Newcastle's neat, tidy, limited war was disproved before the
transports bearing Braddock's two regiments disappeared from
the Irish coast.

Even with the slowness of communications, there were really
no secrets in this age. Spies told the French crown of the sailing;
and in response Louis XV—despite whatever peaceful notions he
might profess to the contrary—ordered some 3,000 of the best-
trained French regulars on the continent to sail for Canada.
Under the command of Jean-Armand, baron de Dieskau, these
troops were to reinforce Quebec and Louisbourg and stand ready
to counter any British advance.

Just as the French knew of Braddock's departure for America, so, too, did the British know about the sailing of these French reinforcements. Nearsighted Newcastle determined to stop the French convoy not in European waters—that might prove too great a provocation—but in American waters, where the remoteness from Europe might lessen the import of the act even if it made finding the French fleet decidedly more problematic. Accordingly, Newcastle dispatched Vice Admiral Edward Boscawen with eleven ships of the line and two frigates to cruise off Louisbourg and "fall upon any French ships of war that shall be attempting to land troops in Nova Scotia or go . . . through the Saint Lawrence to Quebec."[6]

In any encounter on water the odds were that France would almost certainly be weaker. In fact, the French merchant fleet was in such dismal condition that the troops and munitions for these regiments were crowded onto ships of the line instead of transports. Normally carrying sixty-four to seventy-four guns, these battleships were faster than transports, but in order to accommodate the troops, as many as two-thirds of their cannon had to be removed. Therefore, if Boscawen's fleet could catch the French, it would face them with overwhelming firepower.

So, the race was on. The French hoped to sail by mid-April 1755, but hampered by adverse winds, they did not depart Brest until May 3. The fleet was under the command of Rear Admiral Dubois de la Motte, who flew his flag from the seventy-four-gun ship of the line *Entreprenant*, one of only three men-of-war in the fourteen-ship French convoy that had not been stripped down to accommodate troops. Admiral Boscawen had already sailed from Plymouth on April 27, and hoped to be waiting for the French somewhere amid the cold fog of the Grand Banks.

The fog could be friend or foe. De la Motte's fleet arrived off Newfoundland without incident but despite countless signals quickly became hopelessly scattered in the fog banks. Then on June 6, with only three other French ships close at hand, de la Motte's *Entreprenant* caught sight of the bulk of Boscawen's fleet.

Wisely, the French ships slipped back into the murky gloom to avoid it. Three other French ships were not so lucky. The *Alcide*, *Lys*, and *Dauphin Royal* were off the southern coast of Newfoundland west of Cape Race when Boscawen's lookouts caught sight of them and the admiral gave chase.

The sixty-gun *Dunkirk* led the pursuit followed by Boscawen's flagship, *Torbay*, and two other ships of the line. Captain Toussaint Hocquart of the sixty-four-gun *Alcide*, one of the French ships with its armament intact, lay back to cover the flight of the *Lys* and *Dauphin Royal* to the northwest. The former was laden with eight companies of regulars and the latter with nine. As the British ships closed on the *Alcide* with gun ports open, Captain Hocquart was determined not to be the aggressor. If this was Newcastle's limited war, so be it, but France would not be the one to start it. As the *Dunkirk* bore down on the *Alcide*, Hocquart called out in English, "Are we at peace or at war?" When there was no answer, Hocquart hailed a second and third time with the same query, "Are we at peace or at war?"

Finally, the *Dunkirk*'s captain, Richard Howe, replied in French, "*La paix, la paix*"—"At peace, at peace." Scarcely had the words died away when the *Dunkirk*'s cannon belched a broadside at close range. Whether Howe was duplicitous, or Hocquart merely a dupe—there was a red pennant flying from Boscawen's flagship that signaled "attack"—is debatable, but the result was not. *Dunkirk*'s broadside was double-shotted with both cannonballs and grapeshot and chain, and the result was havoc. Hocquart and his crew fought bravely if briefly and soon struck *Alcide*'s colors.

Other British ships closed with *Lys*. Armed with only twenty-four of the lighter cannon of its normal complement of sixty-four, *Lys* ran with a gentle breeze for a time, but was eventually overtaken after a daylong chase. Its captain, too, was forced to surrender or be cut to pieces. Only the third member of the trio, *Dauphin Royal*, which happened to be one of the fastest ships in the French navy, escaped to the safety of the harbor at Louisbourg.

Boscawen was ready to declare a victory, but others were not so sure. One by one, or by twos and threes, the remaining French ships made their way into the Gulf of Saint Lawrence and up the river to Quebec. There was no denying that New France had been reinforced. The loss of the *Alcide* and *Lys* aside, de la Motte had managed to deliver seventy-eight companies of regulars, some 2,600 troops, to the gates of Quebec with the loss of only ten companies and two ships.[7]

"In America, the disputes are," Newcastle had postulated a few months before, "and there they shall remain for us; and there the war may be kept."[8] But in the wake of Admiral Boscawen's actions that was no longer the case. Lord Chancellor Philip Hardwicke, who had participated in the deliberations over sending Boscawen in the first place, put his finger squarely on the problem. "What we have done," Hardwicke wrote to Newcastle, "is either too little or too much. The disappointment gives me great concern."[9] Wherever Edward Braddock's roads led, he was now certain to find forewarned commanders and fresh faces to greet him.

Edward Braddock—the soldier used to giving orders—arrived in Virginia and proceeded to do just that, managing in the process to alienate almost everyone he encountered. Braddock immediately went to Williamsburg to confer with Lieutenant Governor Dinwiddie and then summoned governors De Lancey of New York, Shirley of Massachusetts, Morris of Pennsylvania, and Sharpe of Maryland to meet with them at Alexandria. Rather than ask the governors' cooperation and assistance, Braddock demanded, indeed expected, it. That attitude didn't go over very well with anyone.

"We have a general," wrote William Shirley's son, also named William, "most judiciously chosen for being disqualified for the service he is employed in almost every respect." Assigned to General Braddock as his secretary, the younger Shirley would have cause to feel Braddock's inadequacy all too personally within a few weeks.[10]

Braddock reciprocated this animosity. Having assumed too much the role of the conquering Roman consul, he confessed, "I cannot say as yet they [colonials] have shown the regard . . . that might have been expected." But that hardly kept him from being upset with New York and Massachusetts for continuing to trade with French Canada, indignant at Quaker Pennsylvania for show-ing little enthusiasm for the entire military effort, and decidedly opposed to Governor Shirley's plan to combine efforts and jointly strike Fort Niagara first. That way, Shirley reasoned, Fort Duquesne would wither on the vine like an unharvested grape. Absolutely not, replied Braddock. The duke of Cumberland's orders dictated that the first attack be against Duquesne, and against Duquesne it would be.[11]

Fort Duquesne was certainly not the Gibraltar of the west, but in less than a year's time the French had turned it into a reasonably defensible position and proved young Major Washington's assess-ment of its strategic location. Under the command of Captain Claude-Pierre Pécaudy de Contrecoeur, workers quickly tore down the Ohio Company's hastily begun stockade and in its place erected a solid structure approximately 150 feet square—compact, to be sure, but easily the most imposing fortification on the Ohio frontier. That it should be named for the recent governor general of New France was further proof that the French intended it to be a powerful and permanent fixture.

Squeezed into the very point of the confluence of the Allegheny and Monongahela, the fort had two sides protected by the merging rivers and the other two sides protected by a maze of ditches and embankments. The ramparts facing the land were built of squared logs and backfilled with earth up to ten feet thick. On the watersides, upright round logs formed a stockade wall twelve feet high. Four bastions anchored the corners of the enclo-sure and were topped with an assortment of two- to three-pound swivel guns and six-pound cannons. The main entrance was on the eastern side via a gate and drawbridge, and there was a postern or back gate on the riverside.

Inside the stockade was a minuscule parade ground barely half the size of a tennis court that was surrounded by a guardhouse, the commander's quarters, the officers' apartments, the enlisted men's barracks, and a storeroom. A kitchen, blacksmith shop, powder storeroom, and prison were housed under the four bastions. The forest was cleared for a distance of about a quarter of a mile—more than a musket shot—and the stumps were chopped off at ground level. Corn was planted in this area, where there was also a collection of bark huts and cabins built to house the remaining troops.[12]

General Braddock was quite aware of all this. In fact, he was privy to a very accurate map of the entire French post. His source of information came from Major Robert Stobo, one of two English hostages George Washington had accorded the French after his surrender of Fort Necessity. Sent to Fort Duquesne as leverage to ensure the release of French prisoners captured by the British, Stobo promptly went to work sketching the fortifications and then had his drawings smuggled out of Fort Duquesne by a Delaware Indian.

How much this information may have influenced Braddock on his advance is debatable, but Stobo's sketches of stout ramparts and open approaches may have convinced the general that he was in for a classic siege. In any event, Braddock insisted on dragging with him an artillery train that included monstrous eight-inch howitzers and twelve-pounders taken from a Royal Navy vessel. "I have my own fears," wrote the British admiral from whose ships Braddock procured them, "that the heavy guns must be left on this side of the hill."[13]

Braddock's army of some 2,000 regulars, provincials, and a few sailors assigned to the artillery finally marched out of Fort Cumberland, Maryland, on June 10, 1755—much later than the general had planned. Truth be told, it was lucky to get off at all. Supplies in general and wagons and horses in particular had been painfully difficult to obtain, and only the prompt intervention of none other than Benjamin Franklin had managed to save the day.

Franklin initially contacted Braddock in his role as deputy postmaster general, seeking to facilitate communications among the far-flung prongs of the general's military operations. Quick study that he was, Franklin immediately recognized the muddle with supplies and went to work, going so far as to hint in a newspaper in Pennsylvania that if horses weren't made available for purchase, Braddock might be forced to confiscate some.

Within a couple of weeks, 150 head of horses and wagons loaded with provisions funded in part by the Pennsylvania General Assembly stood ready at Fort Cumberland. Whatever his opinion of colonials, Braddock was deeply grateful for Franklin's prompt leadership in securing Pennsylvania's assistance and acknowledged to Franklin that Virginia and Maryland "had promised everything and performed nothing," while Pennsylvania "had promised nothing and performed everything."[14]

But if at last well provisioned, Braddock's army was nonetheless a lumbering ox when finally under way. Heavy wagons, cumbersome artillery pieces more suited for coastal bombardment than frontier operations, and a string of camp followers all blunted rather than sharpened its strategic focus. Essentially following Washington's route of both 1753 and 1754—which was, of course, really the route of Christopher Gist and his fellow traders—the column took seven days to travel twenty-two miles. It looked as if the journey to the forks of the Ohio was going to be very slow.

Braddock realized this on the evening of June 16 and chose to divide his command. He would push on with approximately 1,300 men, the bulk of the artillery, and about a quarter of the wagons. Reports of increased French reinforcements at the forks were almost as ubiquitous as the trees that were being felled to permit the passage of his wagons, but whatever truth they held suggested that speed was essential. In hindsight, however, all that this division of command accomplished was to weaken a plodding beast, not speed two-thirds to its destination.

Braddock's advance force still had to cut a road through dense forest and over rolling hills, and it even stopped to bridge streams. Another twenty-two days passed before the column arrived on the east bank of the Monongahela some eight miles southeast of Fort

Duquesne. Along the way Braddock repeatedly refused the entreaties of several subordinates that he order up the remainder of his command.

Now, on the evening of July 8, 1755, there was another decision to be made. The route that day had taken the column through a narrow valley lined with steep hills where the column was ripe for an ambush. Braddock's precautions had been textbook perfect: secure the hilltops and mouth of the valley before the main column moved through it. All this had been done without incident, but the route ahead on the east bank of the Monongahela offered a similar passage—dense narrows tightly confined between the river and the high bluffs that spilled onto the ford at Turtle Creek. It was another likely spot for an ambush and this time, on the advice of his guides, Braddock elected to avoid it completely.

Consequently, on the morning of July 9, the column forded

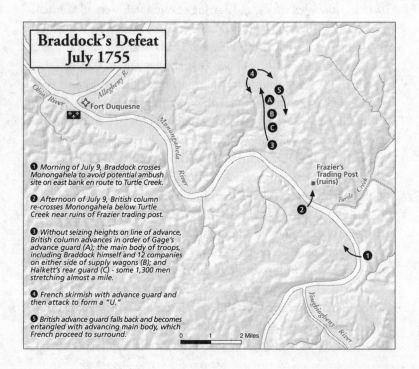

Braddock's Defeat July 1755

❶ Morning of July 9, Braddock crosses Monongahela to avoid potential ambush site on east bank en route to Turtle Creek.

❷ Afternoon of July 9, British column re-crosses Monongahela below Turtle Creek near ruins of Frazier trading post.

❸ Without seizing heights on line of advance, British column advances in order of Gage's advance guard (A); the main body of troops, including Braddock himself and 12 companies on either side of supply wagons (B); and Halkett's rear guard (C) - some 1,300 men stretching almost a mile.

❹ French skirmish with advance guard and then attack to form a "U."

❺ British advance guard falls back and becomes entangled with advancing main body, which French proceed to surround.

Ohio River
Allegheny R.
Fort Duquesne
Monongahela River
Frazier's Trading Post (ruins)
Turtle Creek
Youghiogheny River

0   1   2 Miles

the low waters of the Monongahela to its western bank and proceeded northward through more favorable terrain. It then recrossed the river to arrive back on its eastern bank below Turtle Creek, near the ruins of John Fraser's most recent trading post. (Poor Fraser seems to have been in the thick of things no matter where he hung his hat!)

The order of march for Braddock's column that afternoon— by all accounts a sunny day—was similar to what it had been all along. The Indian agent and trader George Croghan and seven Indian scouts led an advance guard of some 300 regulars under the command of Lieutenant Colonel Thomas Gage. These troops included grenadiers of the Forty-fourth Foot and a company normally commanded by young Captain Horatio Gates, but this day under the leadership of Captain Robert Cholmley with Gates at his side. Behind the advance guard came engineers, axmen, and road builders under the command of Lieutenant Colonel Sir John St. Clair, followed by a long line of wagons loaded with supplies and tools. More troops formed a rear guard to this lead division that "was drawn out over a distance of about one-third of a mile."[15]

About 100 yards to its rear came the main body of troops, headed by a detachment of light horse and another body of workmen, including the British sailors detached to Braddock's service along with the heavy cannon. Next came the general's guard and Braddock himself, accompanied by his two aides-de-camp, George Washington and Robert Orme. Washington was quite ill, suffering from dysentery and hemorrhoids so painful that he "could ride a horse only by tying cushions over the saddle," and he had caught up to Braddock's advance command only the evening before. His absence appears to have been necessitated by his abysmal physical condition, but one wonders if he remembered that the terms of his parole from the French at Fort Necessity the previous year dictated that he not return to the Ohio country for at least a year.

The main force of regulars followed the general and was divided into twelve companies, half on either side of another long line of wagons that carried the bulk of the force's supplies. Finally, there was the rear guard of the entire force, commanded by Sir Peter Halket and including a company of colonial rangers from Virginia led by Adam Stephen, who about thirty years later would passionately persuade Virginia to ratify the Constitution. In all, it was a column of about 1,300 men—reports vary as to the exact number—that stretched almost a mile from Fraser's abandoned post northward into the woods.[16]

Key ingredients clearly missing from this force, however, were Indian allies. The southern Catawba and Cherokee, who might have been inclined to help, remained out of the campaign largely because of a rivalry between South Carolinia's governor James Glen and Virginia's lieutenant governor Dinwiddie. Glen had originally suggested a military conference of southern governors and then championed just the sort of strategy session that Braddock called in Alexandria. But Dinwiddie ignored the first and then blocked Glen's invitation to the latter on petty grounds motivated by political jealousy. Miffed, Glen saw to it that his Catawba and Cherokee allies stayed in the Carolinas.

There wasn't any help from the Iroquois either. At first, the Iroquois were annoyed that the campaign might include their traditional southern enemies, the Catawba and Cherokee. By the time they learned that it wouldn't, those Iroquois who weren't still smarting over the purchases of the Susquehannah Company were equally disgusted with infighting among the British over whether they should join William Shirley's effort against Niagara or William Johnson's against Saint Frédéric. Whatever else they were doing, the Iroquois weren't setting forth for the Monongahela.

Finally, Braddock had held negotiations with groups of Ohio Indians, principally Delaware and Shawnee. The general's bluntness about the purpose of his mission was hardly constructive even to a temporary alliance. Most certainly, Braddock told the assembled chiefs, this was to be the great war for empire—*British* empire, *not* the return of Indian lands from the French. By the time Braddock concluded his oration, these potential allies, too, were offended and quickly melted away into the Pennsylvania woods. That left only seven or eight—depending on whose count one believes—Mingo to scout for Braddock's caravan.[17]

Meanwhile, the French commandant at Fort Duquesne, Captain Pécaudy de Contrecoeur, was not hampered by lack of Indian support. It was, however, the Potawatomi and Ottawa of the western Ohio country, who came to the forks to fight for the fleur-de-lis. According to the historian Francis Jennings, they did so more to curry favor with the French in those western lands than through any great notion of restoring Indian sovereignty east of the Ohio. No matter what their motives, Pécaudy de Contrecoeur would have been in deep trouble without them. By the time the commandant ordered Captain Liénard de Beaujeu, his newly arrived replacement, to lead the bulk of French forces up the Monongahela and intercept Braddock, they numbered 36 French officers, 72 French regulars, 146 Canadian militia, and 637 Indians—roughly two-thirds Braddock's number.[18]

In historical shorthand, the battle that followed is frequently called an ambush. Actually, the initial encounter may have taken both sides by surprise. Beaujeu and his advance units suddenly appeared in front of Gage's lead ranks. Far from being a dense thicket, the forest north of Fraser's cabin was actually fairly open, evidence of an Indian hunting ground that had been cleared of underbrush by burning. The French began to spread out, and Gage's troops fired one or perhaps two musket volleys. At 200 yards, the results were minimal, but among those who fell dead was Captain Beaujeu. Rather than panic, his second in command, Lieutenant Jean-Daniel Dumas, quickly steadied the French regulars across Gage's front, while his Indian allies advanced largely on their own accord to form a U around both sides of the British column. Gage's advance guard retreated and fell back on the first company of wagons and road workers.

For some reason, Gage had failed to send troops to take control of a hill to the right of his line of march, as he had done in similar terrain just the day before. The attacking French and Indians soon occupied this high ground and began to pour a deadly fire into the British column strung out below. Under pressure, Gage's advance guard continued retreating in column, telescoping inward on itself in the process and quickly intermingling the various units into a disorganized mass.

On hearing the sounds of this action, General Braddock's textbook response should have been to order his main body to stand firm until the extent of Gage's engagement could be ascertained. Inexplicably, Braddock instead ordered his main force to advance in column. The result was that the two key pieces of Braddock's command—Gage's advance force retreating down the road and the main body of troops advancing up it—collided in even greater confusion.

The French and Indians continued to fire into this entangled mass of redcoats, who nonetheless stood their ground despite the confusion in their ranks. Braddock and his aides-de-camp, Washington and Orme, were in the thick of things, rallying the troops and directing courageous if somewhat ineffective volley fire into the woods on three sides. Gage's retreat had abandoned two six-pound cannons, and the French now turned them against the head of the British column and added their fire to that coming from the surrounding woods.

By most accounts—and there are many—this situation continued for at least two hours. Then, bullets finally found General Braddock, who had already had five horses shot from beneath him. Braddock fell severely wounded; and as Washington and others moved him toward the rear, the resolve of the British regulars, who—contemporary criticism to the contrary notwithstanding—had valiantly stood their ground, disappeared. Their next maneuver, along with the remainder of Braddock's command, was a headlong rush for the ford at the Monongahela. "When we endeavored to rally them," Washington later wrote, "it was with as much success as if we had attempted to stop the wild bears of the mountains."[19]

So, the flight eastward continued. Four days later, within a mile or so of the ruins of Washington's Fort Necessity, Major

General Edward Braddock succumbed to his wounds. He was buried in an unmarked grave in the middle of the rough road, and the surviving wagons and troops of his command passed over it to obscure its location. Most of the other dead were not so lucky.

Upwards of 500 men from Braddock's command had died. Five years later, their bleached bones would still be visible to a passerby, spread "so thick that one lies on top of another for about a half mile in length, and about one hundred yards in breadth." Among the dead was William Shirley, Jr., shot through the head at the side of the general he despised. Almost as many had been wounded. Some of them would die before reaching Fort Cumberland. By one report, the French losses were twenty-eight killed and about the same number wounded, while the Indian losses were eleven killed and twenty-nine wounded. There was little doubt that Newcastle's plan to secure the British frontier had met with a stunning defeat.[20]

Braddock's defeat is one of those battles that have been fought and refought a thousandfold. In the aftermath of the American Revolution, it was popular to describe Braddock's loss as the defeat of Old World tactics—the opening volley in a contest that ultimately saw the British system unsuited to the demands of America's expansive frontier. But later historians have argued that this interpretation overlooks the fact that the Old World order did not function on the banks of the Monongahela even as it should have functioned in the Old World.

On at least three counts, Braddock's actions that day were counter to Bland's *Treatise of Military Discipline*, the military bible then in use on European battlefields. First, the general had rushed his main body forward into confusion rather than forming up and standing firm to await the report of his advance guard. Second, he had not marched his command by smaller platoon units that could have more readily turned the long column into offensive lines. Finally, Gage had failed to take possession of the high ground above the line of march. And why had Braddock divided his force in the first place? With 2,000 troops arrayed before them, the French might have faced such overwhelming numbers that they would have abandoned Fort Duquesne on their own.[21]

Disastrous as Braddock's defeat was, however, it was to have graver ramifications well beyond the banks of the Monongahela. Among the many wagons and tons of supplies and munitions captured by the French that day was the general's trunk, complete with the war plans of the other three roads of attack, as well as Major Stobo's drawings of Fort Duquesne. This latter evidence proved an embarrassment to Major Stobo, rendering him more a spy than a gentleman prisoner, but the other campaign plans sparked both military maneuvering and political indignation. The French found not only detailed plans for the attacks against Fort Niagara and Fort Saint Frédéric, but also extensive plans to surprise New France and "invade it at a time when, on the faith of the most respectable treaties of peace, it should be safe from any insult."[22] When these documents reached Paris, they were a diplomatic bombshell. So much for Newcastle's limited war.

And what of those other roads that Edward Braddock had contemplated for the summer of 1755? From Albany, Major General William Shirley, who after Braddock's death became commander in chief of British forces in North America, finally marched and paddled 200 miles to Oswego on Lake Ontario, arriving there on August 17, still 150 miles short of Fort Niagara, his goal. As the dog days of late summer took their toll on his men, Shirley received reports of a growing concentration of French troops at Fort Frontenac across Lake Ontario and decided to fortify Oswego and remain there instead. Another of Braddock's roads had been stopped short of its objective.

Albany also saw the departure that summer of Major General William Johnson's force of about 2,000 provincials determined to expel the French from Fort Saint Frédéric. Johnson had a shorter distance to go than Shirley; but just when his advance was moving forward, he stopped to build Fort Edward at the portage between the Hudson River and the southern end of Lake George. He did so, Johnson explained, because "in case they were repulsed (which God forbid) it may serve as a place of retreat."[23]

Meanwhile, Dieskau, having escaped the clutches of Boscawen's fleet, had planned to sail against Shirley at Oswego, but Braddock's captured letters convinced him that he should parry the threat toward Lake Champlain first. "Make all haste," Canada's governor general Vaudreuil urged him, "for when you return we shall send you to Oswego to execute our first design."[24]

Even without Braddock's captured papers, however, the French and Canadians had already been moving south from Fort Saint Frédéric and begun building Fort Carillon at a place the British called Ticonderoga. On September 8, 1755, Dieskau sailed south up Lake George from Carillon. (Lake George drains north into Lake Champlain, which in turn drains north into the Richelieu River and eventually the Saint Lawrence. Thus, sailing south on both lakes is going up the waterway.) Hoping to surprise the British, French regulars along with Canadians and Indian allies crept through the forest and poured volley after volley into Johnson's camp. When the British finally counterattacked, Dieskau was wounded and captured.

Each side claimed victory, but might just as easily have claimed a stalemate. The French continued construction of their fortress of Carillon at the southern end of Lake Champlain. Johnson proceeded to build a second fort, William Henry, at the southern end of Lake George. For the British, this road, too, was a dead end.

Finally, there was Acadia. Perhaps because this prong of Braddock's four-part plan had gotten an earlier start, only here did the British meet with some success. On May 26, 1755, Lieutenant Colonel Robert Monckton sailed from Boston with 2,000 provincials and 280 regulars to attack Fort Beauséjour on the isthmus joining Nova Scotia to New Brunswick. The fort was manned by several companies of French regulars and nearly 1,000 Acadian militia, but when British artillery hauled from Halifax began a bombardment, it was "enough to bring about the surrender of the fort because fire combined with inexperience made everyone in that place give up." Nearby Fort Gaspereau was also captured, effectively cutting the land route between Quebec and Fortress Louisbourg. Capturing Louisbourg would require greater efforts, but in the meantime, the British busied themselves with deporting more than 6,000 Acadians, most to faraway Louisiana, to prevent them from aiding the French cause in either Quebec or Louisbourg.[25]

Notwithstanding Nova Scotia, Braddock's roads in the summer of 1755 had led nowhere, but an interesting chain of subsequent encounters flowed from that day on the banks of the Monongahela and involved three of Braddock's principal lieutenants: George Washington, Thomas Gage, and Horatio Gates. All three earned their spurs of leadership that day, but they would have cause to stand on different sides twenty years hence. By 1774, Thomas Gage was the royal governor of Massachusetts, charged with overseeing a veritable hornet's nest of colonial unrest. His heavy-handed administration only made things worse, and after militia fired on his troops at Lexington, the fat was in the fire. Gage ordered the attack on Bunker Hill and soon found his army besieged by the newly minted Continental Army under the command, of course, of George Washington.

Washington's adjutant general was Horatio Gates, who had gone home to England after the French and Indian War and only recently—with Washington's encouragement—returned to America to settle on land in western Virginia. Within the first two chaotic years of the Revolutionary War, Gates quickly became a major general—again with Washington's blessing—and after the all too common political infighting found himself in command of the American army at Saratoga. The subsequent surrender to him there of General John Burgoyne's British army was arguably of even greater significance than Yorktown. The latter was the finale; but without the decisive victory at Saratoga, the fuse of revolution might have been snuffed out.

By then, Thomas Gage had departed Boston for England and

become the scapegoat of the deteriorating colonial situation generally, and of the British defeat at Bunker Hill specifically. One of those who chastised him was a fellow officer who wrote that Gage was "an officer totally unfitted for this command."[26] That critic was John Burgoyne, who soon would know the ultimate meaning of "scapegoat."

Horatio Gates emerged from Saratoga a hero, but willingly or not soon found himself involved with the "Conway cabal" that questioned Washington's leadership. He was once more at his plantation in Virginia when Congress summoned him to save the Carolinas after the fall of Charleston in 1780, Washington at the time being occupied containing another British army in New York. Gates chose his ground wisely and hoped for another Saratoga, but when most of his militia fled the field, the battle of Camden became one of the most disastrous American defeats of the war. How different, then, were the roads that led from the Monongahela that day in 1755 for Thomas Gage, George Washington, and Horatio Gates.

5

"THAT I CAN
SAVE ENGLAND"

If Braddock's roads did not lead to a strengthening of Great Britain's North American frontier, they did lead to the one place where the duke of Newcastle had not wished to go—all-out war with France. The full-blown conflict was made inevitable not by the frontier struggles in North America, but by a diplomatic upheaval among the powers of Europe. England and France were on opposite sides—there was no question about that—but between 1754 and 1756, the European alliances of the previous war shifted dramatically.

Empress Maria Theresa of Austria started the political shuffle by steadfastly refusing to accept Frederick the Great's accession of Silesia in 1740. Great Britain, Maria Theresa's chief ally in the War of the Austrian Succession, had not proved aggressive enough for her purposes, so the empress went shopping for a new European friend. She found one in her old foe, France, now under the influence of Louis XV's mistress, the marquise de Pompadour.

Once the two women got to plotting, Austria's recapture of Silesia was to be but the first step in dividing up all of Frederick's territory. Empress Elizabeth of Russia, daughter of Peter the Great, eyed Frederick's East Prussia and soon made it a trio against Frederick by allying Russia with Austria and France. Faced with this new alliance, which might well make France the supreme power on the continent, the duke of Newcastle's government had little recourse but to align itself with its old enemy, the enigmatic Frederick.

The remaining European force, which might have tipped the balance of power,* was Spain. Although the days of its glory, before the Armada, were long past, Spain was nevertheless still a key player and the holder of major stakes in the New World. But despite the Bourbon blood of the royal houses in both Spain and France, Ferdinand VI resolutely proclaimed Spain's neutrality. In doing so, he was taking to heart his country's interests much more than the dictate of the Treaty of Utrecht that Spain and France never become one. Much to the chagrin of Louis XV, who relentlessly lobbied for his cousin's involvement, Ferdinand shrewdly recognized that Spain had more to lose than to gain in the conflict. It was a lesson that Ferdinand's successor would have done well to remember a few years later.

So, with new partners, both England and France prepared for global war. Still smarting over Braddock's defeat, Newcastle's England was in no great shape and desperately needed time to build up its forces. France, too, which might have declared war immediately after England's attacks in North America in 1755, was all too happy to have some time to prepare, particularly in building up its navy. Samuel Eliot Morison asserted retrospectively that after the Treaty of Utrecht England was *the* sea power. Three decades after that treaty, however, even England still did not believe it. Lord Anson, soon to hold the Royal Navy's highest post, first lord of the Admiralty, wrote in 1744, "I have never seen or heard . . . that one of our ships, alone and singly opposed to one of the enemy's of equal force has taken her . . . , and yet we are daily boasting of the prowess of our fleet."[1]

But England did have a numerical advantage. In the spring of

* Great Britain's Sir Robert Walpole may have been the first to use the term "balance of power," in a speech in 1741.

1756, according to one estimate of comparative naval strength, England had more than 160 capital ships, including about 100 ships of the line, each armed with 50 to 100 cannons, and more than 60 frigates, each with 32 to 40 cannons. Against this, the French could float only 60 ships of the line and 31 frigates. The number in each class actually fit for service varied greatly with individual reports. But the French were rushing the completion of at least 15 more capital ships either by purchase or in their own yards. Newcastle looked at these figures and feared that should the Spanish Bourbons change their minds, a combined French-Spanish armada would have "decided numerical superiority."[2]

But France was not looking for an epic naval battle in the English Channel. Rather, it hoped to build on the success of de la Motte in eluding Boscawen the previous year. This effort had suggested that France could use its smaller but speedier navy to reinforce North America and its other outposts around the world while keeping England on the lookout for offensive operations closer to home. Eager to adopt these tactics even before a declaration of war, the French minister of the marine, Jean-Baptiste Machault, dispatched three squadrons to North America in the spring of 1756: two to the Caribbean and a third to Canada with another round of reinforcements. Meanwhile, France used its overwhelming strength in land forces to threaten the possibility of an invasion across the Channel.

In England, Newcastle was roundly criticized for fretting about the danger of such an invasion from France. But even though no such attack had occurred since 1066, it was still a possibility— enough, in fact, for King George II to confide that "neither the service of America, neither the defense of Minorca nor any project whatever would incline us to *dégarnir* [strip] our coasts by sending out too many ships of war." Frederick the Great, too, warned the British minister in Berlin of French plans for an invasion. He may well have exaggerated them to solidify his own new alliance with England, but nonetheless, his warning was passed on to Newcastle with the assertion that "we could not be too much on our guard."[3]

• • •

So, with Great Britain thus preoccupied, France decided to make a quick grab for British territory, just as General Braddock had tried to do against French interests the previous year. France looked for a weak spot and found it in the Mediterranean island of Minorca. Thirty miles wide and fifteen miles in diameter, this eastern point in the Balearic Islands off the coast of Spain had a long history of diverse colonization stretching from the Greeks to the Moors. The British had captured it from Spain in 1708, and their ownership was confirmed by the Treaty of Utrecht. Now an important naval base for both trading and military purposes, Minorca was arguably second only to Gibraltar in importance to Great Britain's power in the Mediterranean.

As stealthily as possible, then, the French massed 15,000 men and a train of siege artillery under the duc de Richelieu and sailed out of Toulon aboard 170 vessels on April 10, 1756. Naturally, British spies reported the preparations and Newcastle fretted some more over the fleet's destination. Was it Minorca, Gibraltar, or . . . ? Against the 2,800-man British garrison at Fort Saint Philip on Minorca, this force was excessive, but turned loose elsewhere, perhaps even on the banks of the Thames, it would be of a size to cause utter chaos. Sailing with this convoy were twelve ships of the line commanded by the marquis de la Galissonière, the same man who as governor of New France had dispatched Céloron down the Ohio.

Given Britain's concern over safeguarding the English Channel, its Mediterranean squadron was down to four ships of the line, three frigates, and one lone sloop. Thus it fell to England's second-ranking naval officer, Vice Admiral John Byng, to sail from Portsmouth with a hastily assembled fleet to counter the threat. Byng, a lifelong sailor, was fifty-two and was the son of a storied admiral. He had never been very popular with his crews, perhaps because he was rather austere and a strict disciplinarian. Nonetheless, he had a long career of solid if undistinguished service in the Royal Navy.

By the time Byng's fleet, including his ninety-gun flagship, *Ramillies*, arrived at Gibraltar on May 2, 1756, and joined up with what was left of the Mediterranean squadron, the French forces had landed on Minorca and bottled up its garrison in Fort Saint Philip. It fell to Byng, still operating without a declaration of war, to relieve the siege.

Immediately, there was confusion between Byng's orders and those of the governor of Gibraltar over how many troops should be provided to Byng's fleet. This put Byng in a quandary. He could insist on the troops and sail to relieve the garrison at Fort Saint Philip, but what if Gibraltar was invaded in the interim? Or he could sail with the number of troops he already had, hope to defeat the French at sea, and trust that the garrison could hold out until a larger landing force arrived. Either way, Byng faced potential criticism: fail to save Minorca and be censured; lose Gibraltar and be damned.

Byng chose to hope for a naval victory and sailed from Gibraltar with ten ships of the line and three frigates. Off Minorca, Galissonière was ready for him with what may have been one of the finest naval squadrons ever assembled by France—twelve ships of line that were well built, well armed, well manned, and well trained. In fact, the French flagship, *Foudroyant*, was one of the superweapons of the day, not only mounting eighty-four guns but also including a battery of fifty-two-pounders, clearly capable of throwing more broadside weight than any rival.

Byng attempted to communicate with the beleaguered Saint Philip garrison on Minorca, but on May 20, the sails of the French fleet on the horizon intervened. Having already faced one dilemma at Gibraltar, Byng now faced another. Should he engage a fleet that appeared superior in almost every category, or head for Toulon or other points to draw Galissonière away from Minorca? Galissonière's position was far simpler, and his orders were clear-cut: defend the French beachhead on Minorca and prohibit the reinforcement of its garrison. Byng chose to attack.

Securing the weather gauge, Byng planned to run his fleet in a line obliquely toward the French ships and then tack and turn in line in the opposite direction, continuing to run obliquely at the enemy as the fleets sailed past each other. Properly executed, the plan would have been for British broadsides to pummel the sterns of the French ships, which would have had difficulty turning into the wind to engage their own full broadsides.

It was a good plan, calculated to wreak as much havoc as possible and then permit the fleet to escape intact—ready for the next round. But as the British fleet came about and moved closer to the French ships, the captain of the *Defiance*, now leading Byng's van, misunderstood the signal to sail obliquely and instead led his division straight into the French line. This had the effect of putting British ships' bows on to French broadsides. Byng was furious; and by the time the fleets were disengaged, half of his ships were heavily damaged and no longer fit for action. Byng withdrew, while Galissonière continued his defense of Minorca.

Now what? Byng faced the decision of remaining near Minorca and attempting to land troops despite the French fleet or returning to Gibraltar. After counsel with his officers, it was their unanimous opinion that the fleet should retire to Gibraltar. It was, of course, their opinion but Byng's ultimate responsibility, especially after Fort Saint Philip surrendered a month later.

In a year that held no good news for Newcastle's government from any front, Admiral Byng was subsequently court-martialed, found guilty of numerous counts of misconduct, and sentenced to death by firing squad aboard the *Monarch* in Portsmouth harbor. It was a sentence passed in the heated emotions—both public and governmental—of a year of defeat, and it was not exactly the high point of the Royal Navy.

In retrospect, Byng's concern for Gibraltar and his decision not to risk his entire fleet when other corners of the British Empire were far more dependent on it than Minorca, may well prove his competence. And, of course, if his orders had been carried out competently in the first place, the result may have been far different. Instead, his execution became one of the most egregious affairs in the annals of the Royal Navy. The French philosopher Voltaire, who would pen more than his share of wry comments on the ensuing war, summed the matter up thus: "In this country [England] it is a good thing to kill an admiral from time to time to encourage the others."[4]

Upon learning of the invasion of Minorca, Great Britain declared war on France on May 18, 1756. France reciprocated on June 9. Frederick the Great—without bothering to inform his British ally of his plans—put another log on the blaze by invading neutral Saxony a few months later. The British in North America would come to call the conflict the French and Indian War and trace its beginnings from Washington's defeat at Fort Necessity in 1754. In Europe, where peace and war tended to blur even in the best of times, it would after the fact be called the Seven Years' War. Later still, Samuel Eliot Morison looked at its global scope and the enormous geography at stake and remarked, "This should really have been called the First World War."[5]

Europe itself, North America, and the Mediterranean—where else would the rivals collide? High on the list was India. With the decline of the Mogul Empire in the early 1700s, the subcontinent of India had reverted to a polyglot of largely autonomous provinces under varying degrees of foreign influence. The rich Indian trade that the Portuguese had pioneered before Columbus had long been contested by other European powers, particularly around Bombay on the west coast, Calcutta in the northeast, and Madras on the southeast or Coromandel Coast.

During the War of the Austrian Succession, French forces from their fortress at Pondicherry (in French, Pondichéry) captured the British trading center at Madras and held it until the Treaty of Aix-la-Chapelle. The British responded by building Fort Saint David just south of Pondicherry. In the uneasy peace that followed, the French governor, Joseph-François Dupleix, determined that the French East India Company should use troops left over from the war to acquire territory outright, rather than merely paying local rulers for trading privileges. The English East India Company, assisted by Robert Clive, responded in kind, and the result was that between 1751 and 1754 the two trading companies conducted a limited war all along the Coromandel Coast. England and France were still officially at peace, but—as in the Ohio country during this period—that fact did not stop their respective commercial interests and a handful of regulars on the fringes of their empires from engaging in a struggle to push those limits. The world was becoming smaller.

Dupleix was recalled to France in 1754, and both sides in India momentarily paused to catch their breath. Then, before news of the declarations of war reached India, the new ruler of Bengal, who may have feared the same territorial acquisitions that were taking place around Madras, attacked the British settlement of Calcutta. The British garrison there was sorely understaffed, and the fort's defenses were in a dismal state of neglect. Some Englishmen escaped by ship, but others who surrendered were stuffed into a dungeon that came to be called the "black hole" of Calcutta. Many died of suffocation. France was not directly involved in the infamous affair, but when reports of the "black hole" reached England, it became another sore point for Newcastle's administration and further evidence of global war.[6]

And then came more grave news from North America. It had been a rough year in the wake of Braddock's defeat. Colonel Thomas Dunbar had set the tone early by writing Governor Robert Hunter Morris of Pennsylvania in July 1755 that he intended to retreat from Fort Cumberland with the remainder of Braddock's command and go into winter quarters in Philadelphia. Winter quarters in July! Morris saw red, and so, too, did the Pennsylvania frontier. It was left wide open to marauding companies of French and Indians. One settler reported from the Susquehanna country, "Sure I am, if there is not some speedy measures taken by men of weight, that we shall be utterly ruined." And "in short," the *Pennsylvania Gazette* noted in an editorial, "the distress and confusion our people in general are in on the frontiers is inexpressible."[7]

The situation was no better up north in New York. The *Boston Gazette* had expressed surprise when William Shirley's expedition of

1755 halted at Oswego. "We flattered ourselves," the newspaper noted, "that the three regiments with him, together with the Indians and Rangers, were more than sufficient to have carried both Frontenac and Niagara."[8] By the following summer, however, even Oswego appeared to be out on a limb.

To be sure, Shirley had not forgotten about the two regiments of some 1,500 troops that he had left to quarter Oswego. But after the appointment of John Campbell, the earl of Loudoun, as commander in chief in North America early in 1756, Shirley had less and less to say about them. Before Loudoun arrived in New York, Shirley dispatched Lieutenant Colonel John Bradstreet and a force of about 500 men from Albany to deliver much-needed supplies to Oswego. This Bradstreet accomplished, but on his return journey, French and Indians ambushed his column. Bradstreet held his own, but the attack and his subsequent report of poor conditions at Oswego should have convinced both Shirley and Loudoun that a major relief effort was needed immediately. Unfortunately for the defenders at Oswego, the urgency of the matter fell victim to the change of command between Shirley and Loudoun and frustrating delays of military bureaucracy.

But while the British debated strategy in the aftermath of Loudoun's arrival, the French were on the move in force. Under the command of the marquis de Montcalm, about whom much more would soon be heard, French forces—numbering about 1,300 regulars, 1,700 militia, and assorted Indian allies—sailed south across Lake Ontario from Fort Frontenac and, at Oswego, surrounded the forts of Ontario, Pepperell, and George. Fort Ontario occupied a rise on the eastern bank of the mouth of the Oswego River. It overlooked Fort Pepperell on the western bank and Fort George, a tiny redoubt farther to the west.

Meanwhile, a British relief force under the command of Colonel Daniel Webb finally departed Albany for the west, but it was to be too late. On August 12, 1756, supported by ships in Lake Ontario, Montcalm opened his attack with a bombardment of Fort Ontario by siege artillery. Oswego's commander, Lieutenant

Colonel James F. Mercer, ordered an evacuation of Fort Ontario the next day, but about all that did was to permit Montcalm to place French cannon on the heights and concentrate his fire on the two remaining fortifications. One of the men who fell under this renewed bombardment was Mercer himself, struck dead by a cannonball that beheaded him. His surviving officers held a council of war and decided to surrender.

"I am master of the three forts of Chouegen [Oswego] which I demolish," Montcalm reported, and "of 1,600 prisoners, five flags, one hundred guns, three military chests, victuals for two years, six armed sloops, two hundred bateaux and an astonishing booty made by our Canadians and Indians." Many of the prisoners were shipped to Montreal, but upwards of 100 were massacred in the wake of the surrender before Montcalm could intervene. It should have been a lesson to him for future operations with his Indian allies.[9]

When the news from Oswego spread, British North America shuddered at this crucial defeat. "Oswego is lost," bemoaned a New Yorker, "lost perhaps forever."[10] And with it was lost not only the Lake Ontario fleet but, more important, the gateway to the lucrative fur trade that had made Albany so crucial on the northern frontier. Far from plucking French grapes from the vine that tied Quebec to Louisiana, England had just lost all access to Lake Ontario and much of western New York.

The fiery minister of New England Jonathan Edwards wrote to a friend serving as chaplain for a Massachusetts regiment on Lake George and lamented the loss of Minorca and now Oswego. These defeats, Edwards said, would "tend mightily to animate and encourage the French Nation" and make England "contemptible in the eyes of the nations of Europe. . . . What will become of us, God only knows."[11]

The duke of Newcastle wondered the same thing and looked around for help. Who would save England, if only from itself? Who would save England, if only from its own ineptitude? Not everyone in England believed them, but the answers were forthcoming from a forty-eight-year-old member of Parliament, William Pitt. "I know that I can save England," Pitt vowed, with an assertiveness that had been decidedly lacking in the government, but then also displaying his own enormous ego by quickly adding, "and that no one else can."[12]

William Pitt was born in Saint James Parish, London, on November 15, 1708. History would come to call him the "great commoner," but he was hardly that. His paternal grandfather was Thomas "Diamond" Pitt, a self-made tycoon who had secured the family fortune by trading in India and serving as president of the East India Company and governor of Madras. Diamond Pitt was the original owner of the 410-carat Pitt diamond, dug from the Parteal mines of India and destined to grace the sword of Napoleon. Some said the old man was mad—a claim that would be leveled at his descendants as well—but whether he was truly insane, merely eccentric, or just bullheaded, it was hard to argue with his accomplishments.

"If you ever intend to be great," "Diamond" Pitt advised his son Robert, "you must be first good, and that will bring with it a lasting greatness, and without it, it will be but a bubble blown away with the least blast." As to public service, the old man continued, "if you are in Parliament, show yourself on all occasions a good Englishman, and a faithful servant to your country."[13] It was a charge that Robert took seriously when he did in fact serve in Parliament, and Robert's second son, William, would take it even more seriously.

From these scarcely humble beginnings, William Pitt was dispatched first to Eton and then to Trinity College at Oxford. Ill health—Pitt suffered horribly from gout and other maladies most of his life—kept him from graduating, but he embarked on a self-directed finishing tour on the continent, which included spending time at the University of Utrecht and throughout Holland and France.

Early in 1735, at the age of twenty-seven, Pitt took his seat in Parliament. His journey there had certainly not been that of the average commoner, but he seems to have held a genuine love for liberty and to have realized that it flourished best when the rights of the common man were championed. To the English middle class—still disenfranchised in the mid-eighteenth century—William Pitt became a champion, raising his piercing voice in the House of Commons as their voice. In time, when Pitt spoke, he might not move the government, but the government listened and occasionally trembled nonetheless.

Pitt incurred the wrath of Sir Robert Walpole's government through his affection for King George II's eldest but despised son, Prince Frederick, and in turn used his fiery oratory to help bring Walpole down in 1742, in part over criticism of continuing British subsidies to the king's lands in Hanover. The king, Pitt argued, was turning the nation into "a province to a despicable Electorate [Hanover]."[14] There may well have been great truth in that statement, but it was hardly calculated to win Pitt any favor with the king. So, while Pitt held minor positions, including vice treasurer of Ireland, George II steadfastly refused to draw him into his inner circle.

But in 1754, Newcastle's brother, Henry Pelham, died, leaving a void in the House of Commons. Newcastle himself sat in the House of Lords, and even if he had been inclined toward galvanizing oratory—which he was not—it would have had little impact on the stodgy peerage and even less on the House of Commons. Like it or not, Newcastle looked more and more to Pitt in the House of Commons—out of necessity.

By the fall of 1756, Newcastle's government was reeling under the triple blows of the losses of Minorca, Calcutta, and Oswego. When Henry Fox, then secretary of state for the Southern Department and Newcastle's manager in the House of Commons, abruptly resigned, Newcastle asked Pitt for support, but he did not even find sympathy. Pitt brashly announced that he would not serve in any administration that included Newcastle. The duke saw the handwriting on the wall, made a flurry of last-minute political appointments, and resigned on November 11, 1756. A befuddled George II—who, although there were bouts of insanity in his own family tree, was among those to call William Pitt quite mad—nonetheless reluctantly turned to Pitt to form a government. "I am hand and heart for Pitt, at present," a dejected Newcastle sighed. "He will come as a conqueror. I always dreaded it."[15]

But Pitt was not in for an easy ride. Disdaining the treasury, Pitt chose the duke of Devonshire as first lord of the treasury and the nominal head of his government and himself took the position of secretary of state for the Southern Department. The middle class loved him, but their champion had not exactly endeared himself to the House of Commons at large. Many in Parliament opposed him. The duke of Cumberland did not support him. The king, it was said, loathed him. These were not relationships that would inspire Pitt's confidence or help him to govern decisively. Then, of course, there was still Newcastle. The duke might no longer be the head of the government, but after his nearly four decades in one office or another, the government bureaucracy was filled with his appointees.

Within a month, however, Pitt had boldly written a three-point agenda for George II to deliver to Parliament. First, it recognized the essential importance of the North American colonies to the greater empire. By land and by sea, America must be defended. Second, it created a national militia designed to alleviate fears of a cross-channel invasion, while freeing up regulars for service abroad. Third and last, it called for some measure of relief from the high price of corn and other commodities for the lower class. As with a presidential state of the union address, however, the executive may propose, but how the legislature deposes is an entirely different matter.

It was now up to Pitt to persuade Parliament to craft his ideas into law. He urged that an "expedition of weight" of not less than 8,000 men and a fleet be sent to North America and demanded that the Admiralty provide a list of ships "requisite for the total stagnation and extirpation of the French trade upon the seas." When he found that 62 of Great Britain's 200 warships were out of commission, he began a four-year construction program to bring the Royal Navy up to 400 ships of all classes. But he also crossed both king and citizenry over the execution of Admiral Byng. To his credit, Pitt spoke his mind and his conscience. Whatever the circumstances of that day off Minorca, Pitt thought that Byng's execution was not part and parcel of rectifying them. Against both the crown and popular opinion, Pitt favored clemency.[16]

That was enough for George II. On April 6, 1757, he demanded Pitt's resignation and ordered Newcastle to form an interim government. Almost three months of incessant political bickering and intrigue followed. England drifted without a rudder. The disputes were less about policy—all sides agreed that North America must be saved, and even Pitt now saw that subsidies to Hanover and Prussia were central to keeping France busy on the continent—than about personalities. Who was going to stand at the helm? Finally, it became clear that if neither Newcastle nor Pitt could govern England alone, it could not be governed without both of them.

In June 1757, Lord Chesterfield was instrumental in negotiating a coalition government into which Pitt brought the "confidence and support of the people" and Newcastle brought his far better relations with the king, Parliament, and the bureaucracy. Pitt resumed his office of secretary of state. Newcastle assumed his old post as first lord of the Treasury. In effect, Pitt would be prime minister, left to "appoint generals, admirals, and ambassadors" and to carry on the conduct of the war. Newcastle would be what he was best at being—the wizard behind the curtain working the wheels of patronage, currying favor with the House of Lords, and reassuring the king. "I will borrow the duke's [Newcastle's] majorities to carry on the government," the resurrected Pitt told the duke of Devonshire.[17]

And so a most unusual partnership was born. "The duke of Newcastle and Mr. Pitt," wrote Lord Chesterfield, "jog on like man and wife, that is, seldom agreeing, often quarreling, but by mutual interest . . . not parting." Indeed, their relationship might well be considered the essence of the definition of partnership. Each party needed the other; each brought strengths the other lacked, and each was content—or at least resigned—to let the other do what he did best. "No amount of pressure could create the political machine that was prerequisite for conducting the business of government; the only man with such a machine was Newcastle." And the only man bold enough to use it was William Pitt.[18]

"Britain has long been in labor," Frederick the Great observed later, "and at last she has brought forth a man."[19] It remained to be seen what that man could accomplish. Nevertheless, one fact was now crystal clear. Newcastle had fiddled with it and initially sought to limit its scope, but from now on, there was no doubt but that the conflict in which Great Britain found itself embroiled was Mr. Pitt's global war.

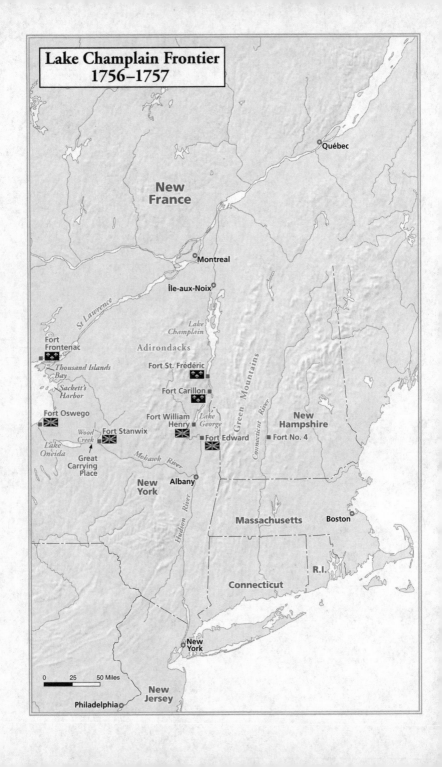

Lake Champlain Frontier
1756–1757

Québec

New
France

Montreal

Île-aux-Noix

St Lawrence

Lake
Champlain

Fort
Frontenac

Thousand Islands
Bay

Adirondacks

Fort St. Frédéric

Sackett's
Harbor

Fort Carillon

Green Mountains

Fort Oswego

Fort William
Henry

Lake
George

Connecticut River

New
Hampshire

Fort Stanwix

Wood
Creek

Fort Edward

Fort No. 4

Lake
Oneida

Great
Carrying
Place

Mohawk River

New
York

Albany

Massachusetts

Boston

Hudson River

R.I.

Connecticut

New
York

0   25   50 Miles

New
Jersey

Philadelphia

BOOK TWO

Mr. Pitt's Global War

(1757–1760)

England and Europe will be fought for in North America.

—WILLIAM PITT
to the Duke of Newcastle, Christmas, 1757

6

MASSACRE AND
STALEMATE

By the spring of 1757 the war in North America—both unde-
clared and declared—had been going very well for the French.
Despite being outnumbered twenty to one in inhabitants, they
continued to hold a vast territory, now more securely protected by
a cordon of forts and Indian allies than ever before. This was a
striking contrast to 1749, when Céloron, on his visit to the Ohio
country, had found the French presence lacking.

In 1754, the French had seized the forks of the Ohio and sent
young George Washington packing back to Virginia. In 1755, save
for the loss of forts Beauséjour and Gaspereau in Nova Scotia,
they had managed to block advances along all of Braddock's roads.
In 1756, they had not only solidified their perimeter defenses from
Fort Duquesne to forts Niagara, Frontenac, Saint Frédéric, and
Carillon, but also swept the British from Oswego and from a toe-
hold on the Great Lakes. Meanwhile, the Indian allies of the
French had played havoc all along the British frontier north of
the Carolinas. With another round of reinforcements in hand and
the fortress of Louisbourg still secure, why shouldn't 1757 be
another banner year for the fleur-de-lis? The answer to that question lay much more in Europe than in North America.[1]

In England, George II had finally resigned himself to William

Pitt's role in his government, although George still grumbled about Pitt's apparent lack of interest in the king's prized possession, Hanover. George and his favored son, the duke of Cumberland, remained transfixed by Hanover's defense among the shifting boundaries of Europe. Pitt, however, painted on a much broader canvas, and it quickly became apparent that this canvas encompassed the world. He had little or no patience with generals and admirals or, for that matter, monarchs, who were content to debate ad infinitum the pros and cons of the push of a single pawn on the global chessboard. Rather, Pitt saw the big picture and demanded bold moves. In truth, the king was right: Pitt cared not a bit for Hanover, but he clearly recognized its strategic importance to keeping France occupied on the continent while he pursued global ambitions.[2]

France took the bait, for a variety of reasons. First of all, the very idea that France might become a great maritime power with a global reach seems to have run counter to its national psyche. England was an island, inexorably tied to the seas around it. France was the crossroads of western Europe, inexorably tied to the lands around it. French armies had held its borders and secured its domain since the days of Charlemagne. Why France could not be a power on both land and sea seems to have been lost on its national consciousness.

Second, the French military establishment, championed by the minister of war, Antoine-René de Voyer d'Argenson, was resolutely determined that the army, not the navy, should be given all priorities and opportunities in *any* military operations. The French army simply could not imagine circumstances in which it would play second fiddle to the navy. Even the successes of de la Motte's resupply of Canada in 1755 and Galissonière's victory off Minorca in 1756 were viewed as tactical support of the army and did not awaken the French to the importance of sea power as a major strategic weapon.

Finally, there was the marquise de Pompadour. Her alliance with Maria Theresa of Austria had turned France's gaze once again to the battlefields of Europe. Rather than emphasize its navy and concentrate its strength against England, its worldwide rival, France chose instead to throw the bulk of its resources into a European land war. After 1757, Louis XV's chessboard was much smaller than William Pitt's, and Louis's overseas colonies, particularly New France, would be neglected. In the words of Frances Parkman, "Louis XV and Pompadour sent a hundred thousand men to fight the battles of Austria, and could spare but twelve hundred to reinforce New France."[3]

With so much glory waiting on the battlefields of Europe, it was no wonder that France's rising military leaders did not fall over each other clamoring for a command in the backwoods of Canada. Louis-Joseph de Montcalm-Gozon, marquis de Montcalm de Saint-Véran, ended up there because he was a good soldier following orders—nothing more. Montcalm was born on February 29, 1712, at the château of Candiac near Nîmes in the south of France. He acquired an early taste for books and received a strict education in Latin, Greek, and history. He joined the French army as an ensign at the age of fifteen, and two years later his father bought him a captaincy. When his father died in 1735, Montcalm was left with an estate that he cherished but that was saddled with debt. Consequently, Montcalm had little choice but to continue his military service. Along the way, he found time to marry and to father ten children.

By 1743, Montcalm had fought in Bohemia and been made colonel of the regiment of Auxerrois. (French regiments were generally named after the province where they were raised, whereas British regiments after 1751 were generally numbered.) More fighting followed in Italy; and in 1746, at the battle of Piacenza on the banks of the Po River, Montcalm twice rallied his troops back from the brink of disaster. In so doing, the marquis received no less than five saber wounds, including two to the head. Clearly, he was not lacking in personal courage. Captured in this engagement, Montcalm was soon paroled to France, where he was made a brigadier.

After another year of warfare, the Treaty of Aix-la-Chapelle gave him an interlude in his fields at Candiac.

In the summer of 1755, the minister of war, Voyer d'Argenson, hinted that Montcalm might be headed for North America, but nothing came of the matter until Voyer d'Argenson looked around for a replacement for the fallen Dieskau after the battle of Lake George. Writing to Montcalm on January 25, 1756, Voyer d'Argenson was all flattery and noted his "greatest pleasure" in announcing that "the King has chosen you to command his troops in North America, and will honor you on your departure with the rank of major-general." Setting out for Paris from his beloved château at Candiac, Montcalm wrote to his mother from Lyons that he had read a pleasant account of Quebec, but "shall always be glad to come home."[4]

Montcalm sailed from Brest on April 3, 1756, with three ships of the line outfitted as transports. On board were two battalions of 600 men each—the 1,200 men Louis XV and the marquise de Pompadour had deigned to spare for the sake of New France. His subordinates were François-Gaston, chevalier de Lévis, named as brigadier; and François-Charles, chevalier de Bourlamaque, named as colonel. Louis-Antoine de Bougainville was to serve as his principal aide-de-camp. After a storm-tossed crossing of the Atlantic, the little fleet anchored in the Saint Lawrence River roughly thirty miles below Quebec, stopped from proceeding farther by heavy spring ice.

Within weeks of his arrival, Montcalm faced two issues that were to weigh heavily on his tenure in North America. The first was his meeting in Montreal with the governor general of New France, Pierre de Rigaud, marquis de Vaudreuil. Born in Canada, Vaudreuil was passionate about its cause, but somewhat indecisive and incompetent in his actions to pursue it. As was the long-established custom, Vaudreuil reported to the minister of marine, while Montcalm reported to the minister of war. Ostensibly, Montcalm was in charge of military matters, while Vaudreuil saw to civilian and administrative ones. But Vaudreuil was less than pleased by Montcalm's arrival;

and the fact that he himself had aspired to both the civilian and the military commands only added to a smoldering distrust between the two men.

The other issue facing Montcalm was the role of France's Indian allies. At Montreal, Montcalm got a quick indoctrination into the intricacies of maintaining their alliances. "One needs the patience of an angel to get on with them," Montcalm wrote to his mother shortly after arriving. "They make war with astounding cruelty, sparing neither men, women, nor children, and take off your scalp very neatly—an operation which generally kills you." Montcalm was to see those tactics firsthand after his capture of the forts at Oswego the following summer, and doubtless they were on his mind as he plotted his strategy for 1757.[5]

But if Montcalm in Canada was feeling a little like the tail of the French dog, the British colonies had yet to feel any warmth from William Pitt's global strategy. In fact, in the spring of 1757, they were feeling quite neglected, and in this they weren't alone among their fellow Englishmen. The *London Evening Post* published a woodcut entitled "Without," depicting the mood in London in the uncertain months between Pitt's first dismissal and his coalition with Newcastle. England, according to this drawing, was feeling "without" about everything from "manufacturers without trade, to colonies without protection, parading fleets without fighting, great armies without use, the common people without money, and the poor without bread."[6]

Given this situation, what were the British colonists to do? They could wallow in the loss of Oswego and wait for Montcalm to renew his attack via avenues from Niagara, Frontenac, and Carillon; or they could launch yet another assault against Canada despite the failures of three years running. To many, those failures proved a point. A jab here or there, or pruning a branch or two off Canada's tree, wasn't going to get the job done. This time, the thrust had to be for the jugular.

Go for Quebec, the *New York Gazette* urged its readers as early as December 6, 1756. Given the logistics of attacking Fort Duquesne or Fort Saint Frédéric, Quebec was no more difficult, the newspaper asserted, and "nothing is more certain than that when the head is lopped off, the inferior members will fall of course; why then is not this effectual step attempted?"[7] Why indeed, but who was to lead it and from where?

In the British chain of command, only one thing was clear. By the end of 1756, William Shirley had become the scapegoat for the fall of Oswego. Not only that, but his successor, the earl of Loudoun, had charged Shirley with near treason. Shirley, Loudoun claimed, had been "raising armies to support himself" by enriching his friends through "lavishing the public treasure" and otherwise impressing people that he was "the only man entrusted in American affairs, by the King or his servants."[8]

Given Shirley's long record of service in North America, it was a low and unjustified blow, but Shirley was summoned home to England to account nonetheless. Loudoun, whom Shirley was to describe as "a pen and ink man whose greatest energies were put forth in getting ready to begin," now had to do just that.[9]

While Lord Loudoun would soon have plenty of his own detractors, he received high marks for his efforts during the fall of 1756 to impose order on the chaotic logistics of the colonial war machine—just the sort of "pen and ink" work at which he excelled. Essentially, this meant improving the stockpiles of clothing, equipment, and provisions for regulars and provincials alike. This required not only the goods themselves, but also a steady and reliable means of moving them from the centralized storehouses at New York, Albany, and Halifax to the principal forts and troops in the field.

Loudoun made progress in this regard, but then ran afoul of provincial governors and their assemblies by instituting an embargo prohibiting all trade between the individual colonies unless it was of a military nature. This was supposed to centralize control over the flow of goods and eliminate the opportunists who were trading with Canada despite the war. About all that it did, however, was to crowd the docks from Boston to Charleston with spoiling produce. By the time that Loudoun realized that the measure was fruitless and lifted the embargo, it had accomplished nothing save harming American commerce. (This was a lesson that Thomas Jefferson would have done well to remember before imposing a far lengthier and far more destructive embargo on the American economy before the War of 1812.)

Given the disaster at Oswego and the impasse around Lake George, Loudoun heartily agreed with the *New York Gazette* and was convinced that the only way to win North America from the French was to strike a blow at its jugular—the Saint Lawrence—and seize Quebec. The question was, by land or by sea? "By fighting across the land," Loudoun wrote to the duke of Argyle, "the troops must be exposed to a thousand accidents . . . and if we have all the success we can hope, we can get no further next campaign than to Lake Champlain." Loudoun's choice was clearly by sea. "By going to Quebec," Loudoun concluded, "success makes us master of everything."[10]

William Pitt certainly didn't disagree with that sentiment, but he did have a different idea about how to attain the goal. England would strike for Quebec, all right, but Pitt refused to do so while his flank was threatened by the fortress of Louisbourg, which, it will be remembered, the English had seized once before—thanks in no small measure to the leadership of William Shirley. Consequently, on February 4, 1757, Pitt dispatched orders to Loudoun to seize Louisbourg before sailing on to Quebec.

But in the winter of 1757, Pitt was in his first tentative ministry and still groping for political power. He needed consensus and was not without critics. The duke of Cumberland in particular favored Loudoun's plan of a single thrust directly against Quebec. Thus, subsequent to Pitt's initial order instructing Loudoun to attack Louisbourg, the consensus of the cabinet broadened his directive to include Quebec at Loudoun's discretion.

This change did not reach Loudoun before he embarked from New York with more than 100 ships carrying 6,000 troops. By the time he received the revised orders in Halifax on July 9, he had also received what proved to be an erroneous report that the French were massing troops at Quebec to repel just the sort of direct thrust he had initially proposed. Despite the discretion allowed in his new orders, Loudoun may well have been relieved to fall back upon Pitt's initial wishes and direct his attention first toward Louisbourg.

Attacking Louisbourg first was certainly the far more conservative step. Fail at Louisbourg, and at least there was hope of a safe retreat to New England. Fail at Quebec with a French garrison and fleet from Louisbourg blocking retreat, and disaster might follow. Loudoun prepared to attack Louisbourg. But meanwhile, his concentration of British troops in Nova Scotia had left forces rather thin along the New York frontier.[11]

Montcalm spent the winter of 1756–1757 entertaining guests in both Montreal and Quebec. Custom and good manners demanded it; adequate though hardly abundant provisions permitted it. Indeed, this winter season may have been the high point of New France. Recent arrivals from old France mingled with Vaudreuil's comrades from the new amid a gaiety and confidence reminiscent of Versailles. From Louisiana to Newfoundland, Canada was secure. The British had been beaten back or contained on all fronts, and Montcalm himself had described his campaign against Oswego as "the most brilliant that has ever been fought on this continent."[12]

Whatever the future held, few could imagine that Louis XV's myopic global vision would be their downfall and that Vaudreuil's own vanity would seal their doom. Throughout the winter, Montcalm wrote endearing letters to his wife and expressed the hope that "next year I may be with you all."[13]

But before that, there was a summer campaign to be fought. Once it became apparent that Loudoun had pulled a goodly number of troops out of the New York area and was heading for Nova Scotia, Montcalm decided to move south from Fort Carillon and attack Fort William Henry at the southern end of Lake George. The fort had already weathered one surprise attack by the French. In mid-March 1757, Governor Vaudreuil's younger brother, François-Pierre Rigaud, led 1,500 French regulars, Canadians, and Indians over the frozen waters of Lake George and besieged William Henry's garrison for four days. Armed only with scaling ladders and lacking siege cannon, the French stood little chance of taking the fort itself, but as the British defenders hunkered down inside, Rigaud's forces were free to pillage and plunder the outbuildings and storehouses, a flotilla of bateaux, and a half-built sloop.

In the aftermath of the French attack, the situation was far graver than might appear from looking only at Fort William Henry's stout walls. Strategically, the fort still guarded the southern end of Lake George and the route south toward Albany, but its effectiveness as a forward base of operations and a stronghold from which to launch an attack against Carillon had been greatly reduced. For the British, the long summer started out with frantic efforts to rebuild some semblance of a flotilla and replenish the fort's supplies.[14]

When Lord Loudoun and his second in command, Major General James Abercromby, sailed for Louisbourg in June of 1757, command of British forces along the New York frontier devolved on Brigadier General Daniel Webb. At first glance, it would seem that Webb should have exuded confidence and inspired it in his men. He had served with the Coldstream Guards, and during the War of the Austrian Succession he had fought at Dettingen in Bavaria in 1743 and Fontenoy in Belgium in 1745. Despite this experience, when faced with the rigors of the New York frontier, Webb seems to have fallen apart. Certainly, he had not inspired anyone when, after the fall of Oswego, he ordered a hasty retreat down the Mohawk Valley.

That Webb remained Lord Loudoun's third in command was not particularly to Loudoun's liking, but Webb held the same advantage as General Braddock—the unwavering patronage of the duke of Cumberland. As Montcalm marshaled his forces and headed south, Webb seemed determined to keep Fort William Henry as a mere pawn pushed to the forefront rather than strengthen it as a bastion designed to control Lake George and threaten Fort Carillon.[15]

Webb's field commander at Fort William Henry was Lieutenant Colonel George Monro. Contrary to the touching story related in James Fenimore Cooper's *The Last of the Mohicans*, Monro had no daughters named Alice and Cora who were determined to rush to their father's side at the beleaguered fort. Monro did have with him five companies of regulars of his Thirty-fifth Foot Regiment—about 600 men—and some 1,200 provincials from New York, Massachusetts, New Jersey, and New Hampshire. But there was serious trouble coming his way. Scouts reported to Monro that Montcalm was gathering upwards of 8,000 men at Fort Carillon.[16]

Colonel John Parker led five companies of New Jersey provincials—the Jersey Blues—north down Lake George to Sabbath Day Point in most of the boats and bateaux that could be mustered to reconnoiter the threat. It was real. Parker's men were surprised off the point by more than 500 Indians—mostly Ottawa, Ojibwa, and Potawatomi—and a few Canadians. The Indians paddled their canoes into the midst of the British flotilla and attacked with a grisly ferociousness, sinking boats and drowning soldiers, many of whom had surrendered. These tactics were hardly new, but they reminded many of the massacre in the aftermath of the siege at Oswego and set a dismal standard for the remainder of the campaign. It was, Montcalm's aide-de-camp Bougainville reported, "a horrible spectacle to European eyes."[17]

Only four boatloads of men escaped the melee and straggled back to Fort William Henry. Their timing could not have been worse. General Webb was at the fort making his first inspection tour. The panic of the returning Jersey Blues was contagious. Completely rattled, Webb promised to send Monro reinforcements and then, just as had happened the year before after Oswego, he scurried in retreat back to Fort Edward.

To his credit, Webb dispatched an additional 200 regulars to Monro's command, but as the French closed in, this was clearly a case of too little too late. Rather than reinforce William Henry with some of the 3,500 men that he had at Fort Edward or with other contingents of provincials stationed at Saratoga and Albany, Webb seemed determined to keep Fort William Henry as a weak outpost.[18]

The rationale of this strategy was completely lost on Colonel Monro. Pleading for reinforcements, Monro urged Webb to fight the decisive battle on Lake George: "If they [the French] are repulsed in their attempt upon Fort William Henry, the affair will be over; but if they take it, I won't say the taking of Fort Edward will be the consequence, but I think it will be a great step toward it."[19]

At dawn on the morning of August 3, 1757, sentries atop the parapets at Fort William Henry reported a fleet of some 250 French bateaux and 150 Indian war canoes bobbing on the waters of Lake George like so many ducks. Some of the bateaux were lashed together with heavy planks and looked like catamarans. It could mean only one thing. "We know that they have cannon," Monro reported to Webb in yet another plea for assistance.[20]

But once again, the French held an advantage over the British more powerful than cannons—their Indian allies. In the two years since Braddock's defeat, the British had not done much to endear their cause to Native Americans. From the Abenaki and Nipissing of Maine westward across the Great Lakes to the Ojibwa, Fox, and Sauk of the upper Mississippi Valley—indeed, southward to the Choctaw along the lower river—most Indian nations could be counted on to support the fleur-de-lis. The two major groups who did not—the Iroquois in the north and the Cherokee in the south—professed a neutrality that wavered depending on who sat before them at their council fires and plied them with presents.

Great Britain's chief Indian agent, Sir William Johnson, certainly tried to persuade the Iroquois to take a major role in Britain's behalf. Lord Loudoun himself strongly supported Johnson's heavy use of presents in pursuit of this goal. But the best that

Johnson could achieve, despite abundant presents, was an uneasy neutrality that meant different things to each side. Even as Montcalm marshaled his forces at Fort Carillon, Johnson told the Six Nations that "to the English their neutrality meant that no French-Indian war parties were to pass through Iroquoian territory and that the Six Nations were immediately to transmit all intelligence to the English."[21] This was wishful thinking.

Everyone likes to be on the winning side, and this was particularly the case with the Indian nations who fought not for a cause or even a continent, but for the loot and plunder of victory. In 1757, the French were clearly winning the war in North America. Theirs was the side to be on. "Our ill success hitherto hath intimidated them," William Johnson reported of his unsuccessful attempts to recruit Indian allies. The British way of warfare was not the Indian way, Johnson continued, and "in short, without some striking success on our side, I believe they will not join us."[22]

But the Indian allies who joined the French at Carillon themselves posed problems for Montcalm. There is no doubt that they were quite an assemblage: by one account, some 2,000 warriors from thirty-three different nations. The French, too, had been busy giving out presents, and these and the promise of victory's rewards had lured some Indians from as far away as the lakes of Minnesota. In modern parlance, command and control in such a situation were bound to be a nightmare, if possible at all. In fact, even as the French recognized that "in the midst of the woods of America one can no more do without them [Indians] than without cavalry in open country," Montcalm also knew that while he might "accommodate, appease, and flatter his allies," he did not truly command them.[23]

So as the sun filled the sky above Fort William Henry on August 3, drums beat a call to arms. Monro and his garrison of less than 2,000 watched as Montcalm's combined force of 7,500 regulars, provincials, and Indian allies unloaded their artillery, quickly encircled the fort, and cut the road leading south to Fort Edward. Help was only fifteen miles away, but with Webb reluctant to provide any, it might as well have been fifteen hundred. "That Fort William-Henry was invested the third instant, by a body of French and Indians," reported the *New York Gazette*, "is past doubt, and very probably are, at this time in possession of it. If so, Fort Edward falls of course, and where they will stop is hard to determine."[24]

Although its outlying works were still showing the results of the French attack of the preceding March, Fort William Henry was nonetheless a position of some consequence. It was built in an irregular square about 130 yards at the widest, and its corners were anchored by four bastions. The northeast bastion stood above the waters of Lake George and the southeastern one above a boggy marsh over which a bridge on the road to Fort Edward now led nowhere. The western bastions were protected by moats topped by a series of palisades. Beyond these lines the trees had been cleared a good distance to afford an advancing enemy no protection. Fort William Henry was certainly not impregnable, but neither was it to be easily brushed aside. The French settled in for a siege and began the backbreaking work of digging zigzagging siege trenches across the open fields. Colonel Monro could do little but choose targets for his cannons wisely and seethe at Webb's outright neglect.[25]

But Webb—for all his recent ineptitude—was in a quandary. If he advanced to Fort William Henry with his remaining force, all might be lost along with the William Henry garrison. If that happened, the road south to Albany would be wide open for Montcalm. So, while Monro pleaded for the decisive battle to be fought at William Henry and not at Fort Edward, Webb sent out frantic appeals of his own for militia from all over New York and New England and hunkered down at Fort Edward. If William Henry fell, Webb might then decide whether to stand and fight at Fort Edward or scurry south to Albany.

Meanwhile, twenty-year-old Sergeant Jabez Fitch, Jr., who traced his family tree back to the *Mayflower*, was serving at Fort Edward in a company of provincials from Norwich, Connecticut. Some of his fellow soldiers wanted to march immediately to the aid of their comrades at William Henry, no matter what the risk. Others, Fitch included, were impatient but content to wait for the arriving militia units because "we are not yet strong enough to engage the enemy." But there was plenty of notice of the contest that was raging just fifteen miles away. "Before sunrise we heard the cannon play very brisk at the lake," Fitch recorded in his diary. "Soon after the small arms began to fire. This firing lasted all day without much ceasing."[26]

But then Monro received a blow more devastating than the thunder of Montcalm's cannon. In a letter to Monro dated August 4, Webb's aide-de-camp had written that General Webb "does not think it prudent . . . to attempt a junction or to assist you" and suggested that Monro consider how he "might make the best terms" of any capitulation.[27] Webb's letter did not reach Monro until August 7. It came under a flag of truce carried from Montcalm by Bougainville after the British courier had been killed in the woods by French scouts. Montcalm added a note of his own, cordially suggesting that Monro take Webb's advice.

Throughout the next day, Monro watched Montcalm's trenches inch closer and closer to the fort. His own effective firepower had been severely reduced over the last several days. Most of his heavy cannons had burst or cracked under the heavy firing. His regulars were anxious, and many of the provincials—totally new to the mental strain of siege warfare—were beside themselves. The commander of one regiment from Massachusetts, Joseph Frye, reported to Monro that "they were quite worn out, and . . . would rather be knocked in the head by the enemy, than stay to perish behind the breastworks."[28]

On the morning of August 9, 1757, Colonel Monro concluded that he had made an honorable defense and in the absence of any support whatsoever from Webb had no option but to surrender. The Union Jack was lowered from William Henry's flagpole and a white flag run up in its place. "We had a rumor about noon that Fort William Henry was taken," Jabez Fitch wrote in his diary, "for their firing ceased some time in the morning."[29]

Colonel Monro dispatched Lieutenant Colonel John Young to Montcalm's camp to negotiate the terms. When agreed to, they were the epitome of the most courteous European warfare of the day. The entire British garrison was to be given safe passage under French guard to Fort Edward. In exchange, all promised that they would not fight again for a period of eighteen months. The conquered would be permitted to keep their personal effects, sidearms, and regimental colors, as well as one six-pound cannon that symbolized their brave defense. Those too gravely wounded to travel would be cared for by the French at Fort William Henry.

For Montcalm's side, in addition to his retaining all the military stores and provisions in the fort, any French prisoners in British custody anywhere in North America were to be returned to Canada by the following November. It was a gentlemen's agreement, quite understood and acceptable to both Monro and Montcalm, but totally foreign to any concept of warfare held by Montcalm's Indian allies.

What happened the following day has many versions, but one undeniable result. Clearly, Montcalm did not "command" his Indian allies, and they had no intention of honoring his promise of safe conduct. The incident seems to have begun when French guards were removed from the British wounded in the fort's hospital. Abenaki fell upon them, murdering and scalping many. The Abenaki were out for the plunder that they had been implicitly promised, and they were determined to get it. Next, they fell to taking items from the British prisoners. How much Montcalm and his officers attempted to restrain them has always been debatable. When French officers suggested that the British give up their baggage and personal effects—despite the terms of surrender—Monro agreed, but this concession had little ameliorating effect on the Indians.

On the morning of August 10, the first large contingent of British troops departed William Henry for Fort Edward, only to be set on by a force of Indians out for much more than just plunder. French guards had been promised, but the few who were present seem to have melted into the woods. "The savages fell upon the rear, killing and scalping," reported Colonel Frye, whose Massachusetts regiment had so desperately wanted to escape the confines of Fort William Henry. They got their wish and indeed were "knocked in the head by the enemy."

To repeat, reports vary greatly, but by August 11 upwards of 700 on the British side had been killed or wounded, or were missing. Belatedly, Montcalm sought the return of those taken captive by the Indians but with limited success. Satiated with the plunder of battle—human and other—Montcalm's Indian allies quickly scattered in the directions from which they had come. On August 15, Colonel Monro and a remaining contingent of soldiers were escorted halfway to Fort Edward by a proper French guard. Incredibly, the survivors of the debacle still chose to drag their six-pounder with them.[30]

What glory there had been in Montcalm's capture of Fort William Henry evaporated on the wind. "Unhappily for the renown of Montcalm and his army," wrote France's greatest military historian of the war, Richard Waddington, "this fine feat of arms was terminated by a horrible massacre."[31] Montcalm's aide, Bougainville, sought to downplay the horror somewhat by suggesting, among other things, that Montcalm had gotten the various chiefs to agree to the terms of surrender. This seems unlikely and quite contrary to the reasons the Indians joined the French campaign in the first place. In fact, Bougainville himself had already come closer to the truth when in the aftermath of Oswego he had characterized Indian warfare as "an abominable kind of war."[32]

What was Montcalm to do now? His Indian allies had gotten what they had come for and were leaving the campaign. Webb cowered at Fort Edward fifteen miles away, but his ranks were swelling daily with the arrival of more and more militia. Montcalm's own militiamen were eager to return home for the harvest. And the British colonies were up in arms. All previous atrocities paled in comparison with the tales that survivors brought back from the woods south of William Henry. Montcalm's Indian allies had broken open a hornet's nest. "Surely if any nation under the heavens was ever provoked to the most rigid severities in the conduct of a war, it is ours!" trumpeted the *New York Mercury*: "Will it not be strictly just and absolutely necessary, from henceforward . . . that we make some severe examples of our inhuman enemies, when they fall into our hands?"[33]

No one would ever suggest that General Webb was the savior of Albany, but in the aftermath of what came to be called the massacre of Fort William Henry, that is what happened. Montcalm looked at the field and, rather than continue southward, decided merely to destroy the remains of Fort William Henry and retire to the security of forts Carillon and Saint Frédéric. Perhaps he remembered Dieskau's misfortune in the woods south of Lake George two years before; but the loss of his Indian allies—and with them much of his ability to gather intelligence—and the unrest of his own militia sealed his decision.

So Albany and even Fort Edward were safe, but the grim affair near Lake George caused tremors despite Montcalm's withdrawal. If William Shirley was still in hot water over the loss of Oswego the previous year, Lord Loudoun—who had been in the vanguard of those urging Shirley's banishment—was suddenly under even greater scrutiny to explain not only the loss of Fort William Henry but also the ensuing massacre. And just where was Loudoun now? Safely ensconced at Louisbourg? Hardly. Loudoun's grand assault on France's Atlantic fortress had been frustrated by the one thing that most of the French military hierarchy seemed to disdain— French sea power.

Yes, it was true. Loudoun's transports had arrived in Halifax on June 30, to be joined ten days later by a Royal Navy squadron boasting seventeen ships of the line under the command of Admiral Francis Holburne. But before they could sail for Louisbourg, no fewer than eighteen French ships of the line and five frigates had congregated in the safety of Louisbourg's fine harbor, waiting for the British to appear.

Holburne spent most of a foul and foggy July trying to ascertain this French naval strength. When at last he did so, he knew that he was at least evenly matched if not clearly outgunned. He undoubtedly also remembered what had happened to Admiral Byng off Minorca. "Considering the strength of the enemy and other circumstances," the admiral reported to Loudoun, "it is my opinion that there is no probability of succeeding in any attempt upon Louisbourg at this advanced season of the year." It was August 4, 1757, the same date that Montcalm's troops had encircled Fort William Henry.[34]

Loudoun and his transports returned to Boston to face only the firestorm of William Henry's collapse, but Holburne's squadron was not so lucky. Reinforced by more ships from England, Holburne determined to lurk outside Louisbourg and pounce on the French fleet as it sailed for Europe in the fall. It was a good plan, but one that failed to account for the fall hurricane season along the Nova Scotia coast.

On September 24, a powerful storm blew in from the southeast and almost drove the bulk of Holburne's fleet onto the rocky coast of Cape Breton Island. The sixty-gun ship of the line *Tilbury* was dashed to pieces on the rocks, and six other ships lost masts to the fury. Most limped into Halifax harbor "in a very shattered condition."[35] Meanwhile, the French fleet rode out the hurricane in the safety of Louisbourg harbor and sailed for France unmolested a few weeks later. Who said that Britannia ruled the waves?

Meanwhile, his majesty's forces were not faring any better on the European continent. After urging Loudoun's direct thrust against Quebec, the duke of Cumberland was dispatched to the continent to save Hanover but instead was surrounded by the French at Hastenbeck in Germany. The duke made his personal escape only through the bravery of his aide-de-camp, Colonel Jeffery Amherst, in directing a rearguard action. Soon afterward, Cumberland was compelled to sign the Convention of Kloster-Zeven, effectively surrendering Hanover to the French. George II's most treasured son had lost his father's most treasured possession. The duke of Cumberland resigned his military offices in disgrace. The only winner in the dismal affair was William Pitt, who saw his hand in military matters considerably strengthened.[36]

For a time in 1757, Great Britain's European ally, Frederick of Prussia, did not fare much better. Defeated in Bohemia by the Austrians, Frederick suffered raids on Berlin by the Russians and Austrians and a French occupation of Hanover. Then with winter approaching, just as Frederick appeared down for the count, he struck a quick one-two punch that gave credence to his appellation "the Great." He handily defeated the French at Rossbach in Saxony and then crushed the Austrians at Leuthen (Lutynia) in Silesia.

Frederick was certainly not out of the woods, but although the Austrians recovered, the French never would. They had been thwarted in their grand scheme to seize Europe's balance of power. "Rossbach," Napoleon said years later when he himself attempted to seize that balance, "was the battle that started the Bourbon regime toward its collapse in revolution."[37]

In the meantime, their defeat at Rossbach meant that the French would become even more transfixed by European battlefields at the expense of North America in the years ahead. Faced with the stigma of a gruesome massacre, the best the French could hope for at the close of a year that had dawned so brightly was a stalemate. The marquise de Pompadour was even more succinct after Rossbach. "After us the deluge," bemoaned the would-be empress whose shortsighted pettiness in trying to win one continent almost assuredly cost her another.[38]

7

FORTRESS ATLANTIS

Suggestions of a stalemate aside, as 1757 drew to a close it still appeared to many observers of the New World—French and British alike—that the war in North America was being decisively won by the French. Not only did Montcalm write to his superiors anticipating its quick end, but he boasted of French claims as expansive as ever. He proposed that should time permit between a peace treaty and his return to France, he would very much like to survey the Great Lakes and the upper Ohio "with military and political views."[1]

Clearly, Montcalm thought that New France was expanding, not contracting. But the festering political problem for the French in Canada in the new year of 1758 was that Old World observers in France saw the entire North American continent as a disjointed and most unwelcome distraction from European battlefields. William Pitt, of course, saw just the opposite.

With the duke of Cumberland cowed by his own defeat in Europe, Pitt was now free to accelerate his global strategy. Central to implementing this strategy was his firm belief that "Canada was to be attacked from all sides and exhausted." To do so, Pitt devised a massive three-prong offensive. The first prong was to be directed against the citadel of Louisbourg. Lord Loudoun's effort there in 1757 had never hit the beach, and this time Pitt was determined to staff the attack with far more aggressive leaders.

The second prong was to stab northward from the muddle at

Fort Edward and take the French positions at Forts Carillon and Saint Frédéric. With that done, the way down Lake Champlain and the Richelieu River to Montreal would be open.

Finally, Pitt's third great prong was to revisit General Braddock's ill-traveled road, seize Fort Duquesne, and then swing northward to Fort Niagara, effectively cutting ties between Canada and Louisiana. To make this all happen, Pitt tenaciously committed 20,000 British regulars and 22,000 colonials to the coming campaigns, "numbers such as had never before been thrown into the balance of colonial conflicts."[2]

After the massacre at Fort William Henry, the *New York Mercury* avowed as how "'tis certain that the growth of the British colonies has long been the grand object of French envy; and 'tis said that their officers have orders . . . to make the present war as bloody and destructive as possible!"[3] The French had indeed done that, and the result was that even the Quakers of Pennsylvania— besieged by fierce Indian raids all along their western borders— had come to heed Benjamin Franklin's admonishment that "it was useless to hope for a permanent peace so long as the French were masters of Canada."[4] In fact, thanks in large part to these French and Indian pressures, the British colonies were beginning to cooperate in ways that might have surprised even the delegates at Albany four years before.

But any military and political concerns overlooked a fatal weakness in Canada that was even more acute than Louis XV's lack of interest. New France was starving. Never mind the glories or, in the case of William Henry, the indiscretions on the battlefields. What good were they if the colony could not feed itself? With a limited agricultural base, provisions from France had never been more than barely adequate and these slowed to a trickle as Pitt tightened the Royal Navy's noose around the mouth of the Saint Lawrence.

There perhaps was no more graphic example of the fundamental difference between New France and the British colonies than two scenes played out in September 1757. That autumn, despite the horror and havoc recently visited on their compatriots at Fort William Henry, British colonists at King's College in New York celebrated the start of the academic year with abundant pomp and ceremony. The frontier might be in disarray, but the underpinnings were secure. Life went on. Meanwhile, about 450 miles to the north, the Quebec Seminary took the extraordinary step of dismissing its students because there was not enough food to feed them. It was an enormous contrast and would only grow more so, as Pitt's armies sought to encircle Canada.[5]

But to whom would Pitt turn to do his military bidding? His disdain for many in Great Britain's high command was well founded. After all, given recent events both in North America and on the continent, there were certainly no shining stars. Pitt might well be crazy—as some critics charged—but history would also come to call him a shrewd judge of men. In the first of many such appointments, Pitt reached down through the royal ranks and elevated a promising young colonel to command the Louisbourg prong of his grand scheme.

At first glance, Jeffery Amherst was an unlikely choice. Unquestionably, he had been a loyal and competent aide to three notable patrons, but he had never held center stage himself or, for that matter, led an independent command. Amherst had been born in Kent on January 29, 1717. His father was another Jeffery Amherst; and, as the second son, young Jeffery was destined for the military. His first patron was his father's neighbor the duke of Dorset. Jeffery served Dorset as a page, and the duke bought him an ensign's commission before he was fifteen. Later, on Dorset's recommendation, General Sir John Ligonier made Amherst his aide-de-camp. After duty in Flanders and at the battles of Dettingen and Fontenoy, Ligonier recommended Amherst to the duke of Cumberland. By now forty years of age and a lieutenant colonel, Amherst served on Cumberland's staff long enough to earn the duke's admiration, but also to suffer the indignities of his defeat at Hastenbeck and the subsequent humiliation in the Convention of Kloster-Zeven.

Jeffery Amherst's greatest patron, however, was to be William

Pitt, thanks in no small measure to the continued recommendations of Ligonier and to kind words as well from the duke of Cumberland. The latter was now disgraced, but the former had become his replacement as commander in chief of the British army. So it was that in the icy cold of January 1758, in Stade in northern Hanover, Colonel Jeffery Amherst received instructions to report posthaste to London. He did so as soon as ice in the Elbe River permitted his ship to sail and was soon afterward commissioned "Major General of Our Forces in North America, and Commander-in-Chief of a Body of Our Land Forces, to be employed in the siege of Louisbourg."[6]

To some, Louisbourg on Cape Breton Island was an impregnable fortress—France's North American Gibraltar before the real Gibraltar had attained such significance for the British. Indeed, the mere mention of French activities at Louisbourg made most of New England tremble. For a time in the summer of 1755, Boston feared that an invasion from Louisbourg might be imminent. Then, the *Boston Evening Post* printed a correction of a slipped digit. "Our candid readers are desired to correct or pardon an error in our last Monday's paper," pleaded the *Post*, "and instead of 13000 French troops said to be arrived in Louisbourg, read 1300."[7] But there were still plenty of British colonists who groused that the crown should never have returned the fortress as part of the peace of Aix-la-Chapelle.

It will be remembered that under the terms of the Treaty of Utrecht at the end of the War of the Spanish Succession, Great Britain managed to secure its first Canadian toeholds in an attempt to constrain New France. Great Britain received Newfoundland, whose remoteness kept it out of the story; and mainland Nova Scotia, whose proximity to land retained by the French—Ile Royale (Cape Breton Island) and Ile Saint Jean (Prince Edward Island)—had just the opposite effect, setting up half a century of border maneuverings and warfare.

Starting out as a small base for cod fishing, Louisbourg quickly became the linchpin of these French possessions as well as the guardian of the sea lanes into the mouth of the Saint Lawrence River. By the 1730s, more than 150 ships called at Louisbourg annually, making it one of the busiest seaports in North America. A decade later, its permanent population had swelled to almost 3,000. While many inhabitants were French, Louisbourg was a cosmopolitan place and counted several hundred Basques and 150 Germans and Swiss among its inhabitants. William Pepperell's capture of Louisbourg in 1745 stirred things up for a while, but when the dust of the Treaty of Aix-la-Chapelle settled, Louisbourg was again in French hands.

The phrase "impregnable fortress" suggests a high-walled castle bristling with cannon atop a rocky precipice, but Louisbourg was hardly that. Rather, it was a fortified town whose main defenses stretched nearly a mile. Not only did the generally flat and marshy terrain preclude any grand hilltop edifice, but eighteenth-century improvements in artillery had long rendered such medieval castles obsolete. Instead, the design of Louisbourg's main fortifications relied on stout walls banked on the attacking side with earthworks, or glacis, to deflect or absorb cannon fire, and backed on the defenders' side with earthen ramparts topped with parapets from which artillery and small-arms fire could be returned. These walls, called curtain walls, were anchored by multisided bastions from which fire could be directed in several directions and used to protect and cover other bastions as well as the connecting curtain walls.

Seven major bastions encircled Louisbourg. Four protected the landward side, chief of which was the King's Bastion. Three others secured the eastern approaches from the harbor and Rochefort Point. No one could enter the town except through one of several heavily guarded gates; the principal one was the Dauphin Gate below the Dauphin Bastion adjacent to the western harbor.

The harbor itself was defended by four major artillery batteries, two of which were located in the bastions on either side of the waterfront. A third battery occupied the small island at the mouth of the harbor. The fourth sat on the north shore, from which its field of fire could sweep the entire harbor. In the summer of 1758, some 3,500 French soldiers and militia under the command of Governor Augustin de Drucour manned these fortifications and waited for what they knew was inevitable.[8]

Indeed, what the British were up to was certainly no secret. Once again, French spies had only to look at English newspapers and sort through a little misinformation to read the obvious. In North America, the *Pennsylvania Gazette* boasted that the British fleet sailing for Louisbourg would be the "greatest, best manned, and otherwise the best equipped of any fleet that sailed from England since the last Dutch War."[9] This was debatable, of course, but what was not was the fact that the British were to have the North Atlantic increasingly to themselves.

Admiral de la Motte's French fleet, which had proved Louis XV's high command wrong on its two previous voyages to North America, had returned to France in the fall of 1757 with its crews decimated by disease and its ships badly in need of repairs. Other French naval units had fallen victim to British victories in the Mediterranean, and a French resupply fleet assembling near La Rochelle on the Bay of Biscay was soon scattered.

Only Captain d'Escadre des Gouttes aboard *Prudent* managed to slip out of Brest harbor with four other ships of the line and a frigate and make for the open North Atlantic. Later, another two ships of the line and four others converted to transports did likewise, but only about half of these little fleets made it safely into Louisbourg's harbor. Once there, they faced entrapment and posed little strategic threat to the converging British navy.

Meanwhile, Admiral Edward Boscawen departed England on February 19 with ten ships of the line and an assortment of frigates. Numerous transports carrying two regiments and supplies had already sailed for Halifax. With Boscawen in his fleet was one of Jeffery Amherst's chosen lieutenants, Brigadier General James Wolfe. Two other brigadiers—Edward Whitmore and Nova Scotia's governor, Charles Lawrence—were to rendezvous with them in Halifax with additional ships and forces of both regulars and colonials.

Pitt's orders were that once 8,000 troops assembled at Halifax, they were to embark for Louisbourg under Lawrence's command—with or without their commander in chief. General Amherst was detained in England until March 16; and Pitt, it seems, was most anxious that the Louisbourg campaign should begin early enough in the season that a subsequent advance up the Saint Lawrence against Quebec might be possible before fall.[10]

Amherst's general orders were quite similar to those already given to Admiral Boscawen. With both his newly appointed major general and the veteran admiral, Pitt went to great lengths to stress the need for cooperation with the other service—something far more rare than might be expected. In this endeavor, Amherst and Boscawen were to form a fruitful partnership. The amphibious landings on Cape Breton Island—Boscawen, of course, was particularly well versed in the specifics of Louisbourg, not to mention the fickle weather that swirled around it—and their accompanying naval operations were to set the standard for later such operations.

"What an amazing change over the preceding years," wrote the historian Walter Dorn, with no small amount of hyperbole, "to find the army and navy, Englishmen and colonists, co-operating toward a common end and inspired by the spirit of heroic enterprise! Henceforth Canada was a beleaguered fortress."[11]

But if General Amherst was to be delayed, so too were Boscawen and Wolfe. The admiral's ships took a miserable eleven weeks to cross the storm-tossed Atlantic, and most, including his flagship, the ninety-gun ship of the line *Namur*, did not reach Halifax Harbor until May 9, 1758. Perhaps no one in the entire fleet was more seasick than Brigadier General James Wolfe, who weathered the crossing on the eighty-gun *Princess Amelia*. "From Christopher Columbus's time to our days there perhaps has never been a more extraordinary voyage," wrote Wolfe, with shaky pen. "The continual opposition of contrary winds, calms, or currents baffled all our skill and wore out all our patience."[12]

This was putting it mildly. Patience was not all that had worn thin for Wolfe. He was tall and slight—one might say gangly—with reddish hair and a constitution given to a host of chronic ailments. He had been born in Westerham, Kent, on January 2, 1727, not very far from the birthplace of Jeffery Amherst, who was ten years his senior. Wolfe's father, Edward, had served under Marlborough in the Netherlands and Scotland and risen to the rank of major general. In 1741, at the age of fourteen, young Wolfe was given a commission as a second lieutenant in his father's regiment.

Two years later, at Dettingen in Bavaria, Wolfe, though only sixteen, received his first real test in battle. His regiment stood in the center of the British line and took the most casualties. Wolfe reported to his father that he was acting adjutant and had his horse shot from underneath him so that "I was obliged to do the duty of an adjutant all that and the next day on foot, in a pair of heavy boots."[13] Two years after that, at Culloden against the last gasp of the Stuarts, his regiment again suffered the most, losing one-third of its men. Clearly, young Wolfe was not one to run from the sound of the guns.

This rather frail man quickly achieved a reputation as a competent regimental officer that seems to have baffled even him. "I reckon it a very great misfortune to this country," Wolfe wrote to his mother in 1755, "that I, your son, who have, I know, but a very modest capacity, and some degree of diligence a little above the ordinary run, should be thought, as I generally am, one of the best officers of my rank in the service."[14]

His humility did not, however, keep Wolfe from being peeved when he was twice passed over for the command of the Twentieth Foot Regiment—no doubt because some judged him too young. In the meantime, a month past his thirtieth birthday, Wolfe was made quartermaster of troops in Ireland. He found the duty less than stimulating; and when both battalions of the Twentieth Foot were deployed for a raid against Rochefort on the French coast, Wolfe was all too eager to join them, this time as quartermaster of the entire expeditionary force.

The raid at Rochefort was conceived by Pitt as a way to put pressure on France's western borders and thus offer some relief to his Prussian allies without actually committing British troops to the continent. It proved a boondoggle from beginning to end. About 8,400 British soldiers and marines embarked in fifty-five transports protected by thirty warships. The British occupied the Ile d'Aix, but when junior officers, including Wolfe, urged an advance on the mainland as planned, the council of staid British admirals and generals commanding the force balked and ordered a retreat.

"We blundered most egregiously on all sides—sea and land," Wolfe wrote to his father, but he later confessed that he was "not sorry that I went, notwithstanding what has happened; one may always pick up something useful from amongst the most fatal errors." One thing Wolfe seems to have learned is that any amphibious landing must not dawdle, but rather strike quickly and "lose no time in getting the troops on shore."[15]

Pitt suffered much embarrassment over this episode, but he stoutly weathered the inevitable criticism, believing that the failure of the expedition was born of poor execution, not a poor concept. Contrasting Wolfe's role with that of his superiors, Pitt predictably marked Wolfe as a man to get things done. Three months later, Wolfe was summoned to London and offered a promotion to brigadier general and command of a brigade in the Louisbourg expedition. He accepted, of course, but acknowledged that "the very passage [crossing the Atlantic] threatens my life, and that my constitution must be utterly ruined and undone."[16] Whatever else he was, James Wolfe was definitely not a sailor.

On May 28, 1758, following Pitt's instructions not to wait for General Amherst, Boscawen's fleet sailed out of Halifax bound for Louisbourg with brigadiers Lawrence, Wolfe, and Whitmore onboard. The approximately 150 warships and transports had barely cleared the harbor, however, when the sails of a lone British ship of the line appeared on the eastern horizon. It proved to be

Captain George Rodney's seventy-four-gun *Dublin*, but it flew the pennant of Major General Jeffery Amherst. The commander in chief had arrived in time after all—barely.

Though a far better sailor than Wolfe, Amherst, too, had had a long and tedious crossing, interrupted by Rodney's diverting to capture a French merchantman off the coast of Spain. But now, after ten weeks of worrying in his journal, Amherst reported almost casually that "we saw the land off Halifax Harbor and about eight o'clock saw several ships coming out."[17]

While Lawrence, Wolfe, and Whitmore welcomed their commander warmly, one can only wonder what their true feelings were. Suddenly, they were no longer on their own. Before Amherst's arrival, Wolfe and his fellow brigadiers had devised a three-prong attack. Wolfe was to lead the main force in a landing at Miré Bay, north of Louisbourg, and then move southward to join up with troops who would come ashore on the southern coast and move north, effectively encircling the landward approaches to the town. Meanwhile, Brigadier General Charles Lawrence's division and Admiral Boscawen were to make two separate diversions and confuse the French as to the true landing points.

Amherst liked the idea of a diversion and had no qualms about entrusting Wolfe with the main attack, but he insisted that all operations be concentrated in Gabarus Bay on the southern shore where Pepperell's New England troops had successfully landed in 1745. Accordingly, Boscawen's fleet anchored in Gabarus Bay; and on June 2, 1758, Wolfe, seasick as usual, accompanied Amherst as he reconnoitered its shoreline.

Initially, Amherst favored a coordinated attack by three divisions along the full extent of Gabarus Bay. Wolfe was to move left of Flat Point against what the British called Kennington Cove (the French called it Anse de la Cormorandière, Cormorant Bay); Lawrence would attack in the center between Flat Point and White Point; and Whitmore would circle around White Point and come ashore to the east of it. But as the high tides of late spring pounded the rocky coastline and a heavy fog ensnared the fleet, more than one British soldier wondered if he would ever get to land. Doubtless none was more anxious to do so than Wolfe himself.

By the time the heavy swells had subsided somewhat, it was the early hours of June 8 and Amherst had altered his plan. Wolfe and his division were now to force the way into Kennington Cove on their own, while the divisions of Lawrence and Whitmore were to create diversions and then either move ashore as originally planned or row left to support Wolfe.

The French, of course, were hardly oblivious to any of this. The preceding year, Admiral de la Motte had employed large numbers of his sailors to aid the French garrison in constructing a series of earthworks at every feasible landing site along the coast. Large trees were cut and placed in front of the works, their limbs forming an abatis-like tangle. Behind these fortifications, from Kennington Cove eastward to Flat Point and around White Point, there were now posted some 2,000 regulars, militia, and Micmac Indian allies under the command of Colonel St. Julien. Artillery pieces were interspersed among them like coiled rattlesnakes ready to strike.

At the appointed hour, Wolfe led his division into this waiting maelstrom, feinting first toward Flat Point and then angling left into Kennington Cove as planned. The first enemy his troops encountered was the pounding surf, which capsized some boats and splintered others on the gnawing rocks. Then, as the British boats struggled closer toward shore, the French artillery pieces belched their destructive missiles and French musket fire added its staccato to the din. Wolfe looked about him and quickly recognized the obvious. His force was in danger of being cut to pieces.

Standing erect in the pitching bow of his boat, Wolfe raised his hat high over his head and signaled furiously. But what did he mean? Some historians have argued that Wolfe was desperately trying to call off the attack. Perhaps. But at that same instant, a boatload of Highlanders saw a partially protected sandy nook in the rocky shore and rowed toward it with a vengeance. Other boats quickly followed, and soon Wolfe himself was leaping into

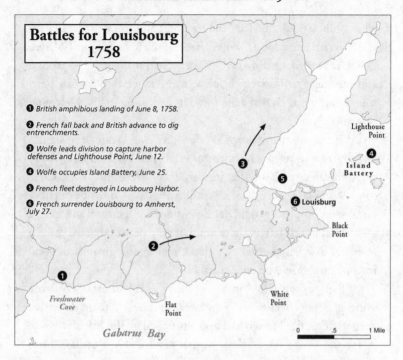

Battles for Louisbourg 1758

❶ British amphibious landing of June 8, 1758.

❷ French fall back and British advance to dig entrenchments.

❸ Wolfe leads division to capture harbor defenses and Lighthouse Point, June 12.

❹ Wolfe occupies Island Battery, June 25.

❺ French fleet destroyed in Louisbourg Harbor.

❻ French surrender Louisbourg to Amherst, July 27.

Lighthouse Point

Island Battery

Louisburg

Black Point

Freshwater Cove

Flat Point

White Point

Gabarus Bay

0   .5   1 Mile

the knee-deep surf. Armed only with his cane, the lanky brigadier urged his troops onward.

And onward they came. As company after company struggled ashore, Wolfe formed them into ranks in the face of French fire that was even more withering than it had been in the boats. But once the British troops returned volley fire and began to move inland, the thin French line collapsed. Soon, not only Wolfe's entire division but the divisions of Lawrence and Whitmore as well were streaming ashore, all concentrating in this one area of Kennington Cove. The French remained spread out along the entire shore of Gabarus Bay until it was too late. Those French troops near Wolfe's landing site feared being encircled and beat a hasty retreat toward Louisbourg. Soon, all of St. Julien's remaining troops around Gabarus Bay were following them.[18]

By nightfall, Amherst had joined Wolfe onshore and advanced with his troops to within range of the guns of the city itself over what Amherst described as "the roughest and worst ground I ever saw."[19] With one wave of Wolfe's hat and the loss of about 100 men killed, wounded, or drowned, Amherst had managed to surround and lay siege to Fortress Louisbourg. Amherst's plan had been sound, but it was clear that he owed its execution to his able young brigadier.

What about that wave of Wolfe's hat? Had Wolfe been signaling a retreat or urging his men forward? For his own part, Wolfe was forever coy about the answer, acknowledging to his uncle some weeks later only that "it is impossible to go into any detail of our operations: they would neither amuse nor instruct, and we are all hurried in our letters. In general, it may be said that we made a rash and ill-advised attempt to land, and by the greatest of good fortune imaginable we succeeded."[20]

Other reports filled in a few more details, but reached the same general conclusion. Somehow, despite all their defensive preparations, the French had been surprised by the British resolve in forcing a landing. Indeed, British resolve seems to have been met with only halfhearted resistance by the defenders.

"Knowing your eyes are turned on us, and that you are impatient to hear how we go on," wrote a British observer the day after the landings, "I take the first opportunity to inform you, that our troops are all landed in front of the enemy who had fortified every place where they thought it practicable for us to land. It was a place not to be forced if those [who] were to defend it had done their duty."[21]

And duty, or a lapse of it, seems to have been at the core of the French defeat. Most of the French line along Gabarus Bay was preoccupied by the diversions of Whitmore and Lawrence's divisions rowing about. No one seems to have recognized the seriousness of Wolfe's initial landing until, in the words of one French report, "several thousand of English soldiers had been landed and drawn up in battle array, having cut off the regiment of Artois [about 500 men] from the rest of our troops."[22]

Once the Artois regiment broke and fled toward Louisbourg,

the remaining French troops did likewise as Wolfe's troops pressed on eastward along the coast and until "all Gabarus Bay was ours by one o'clock." A captured French grenadier captain described the attack as "desperate and presumptuous," and said that "no people in the world but the British troops would have attempted and carried it." Perhaps, but an English commentary published in Boston recognized that while "no people behaved better than our troops, and more cowardly than the French," the French troops were "very sickly for want of fresh provisions" and anxious to desert if given the opportunity.[23]

There it was again—the issue of food. How could troops of any nation be expected to fight if their stomachs were empty and had frequently been empty for much of the preceding winter? Since becoming governor of Louisbourg in 1754, Drucour had pleaded incessantly with the French government in Paris for two things: more troops and more provisions. Drucour, the younger son of a noble Norman family, had entered the French navy in 1719 and served in a variety of posts. As his responsibilities increased, he had been almost ruined financially by the entertainment he was expected to provide because he "preferred to maintain the dignity of any position to which his sovereign had called him, rather than exercise a reasonable regard for his private interests."[24]

There was no doubt that Drucour was determined to maintain the dignity of his position at Louisbourg, but rarely had his pleas for troops and provisions been answered in sufficient quantities. This proved increasingly true as British naval superiority increased and reprovisioning Louisbourg became more and more of a gamble. The winter of 1756–1757 had been particularly bleak and "not a family had an ounce of flour in the house" by the end of a winter that was so long that "there remained eighteen inches of snow on the ground on the twelfth of May." Only the arrival of a lone supply ship in January 1758, far later in the season than many thought possible, had sustained the garrison through yet another cold winter. Now, with Amherst's army busy digging siege trenches just beyond his walls, what was Drucour to do?[25]

As mentioned above, a scattering of French ships had managed to elude the British fleet earlier in the spring and reach Louisbourg. Now, their importance as a means of delaying the British came into play. Under a more aggressive commander, they might have emerged from Louisbourg Harbor to contest the British landings in a daring raid. Indeed, at this season of the year the prevailing easterly winds would have favored the French ships leaving Louisbourg Harbor. Taking advantage of the frequent fogs, they might have run westward with the wind in an hour's time and surprised the British fleet that was crowded into Gabarus Bay. The prevailing winds would have made maneuvering away from the rocky shore difficult for the British. The French ships, of which there appear to have been at least five ships of the line, including the seventy-four-gun *Prudent* and seventy-four-gun *Entreprenant*, might have been able to inflict considerable chaos, if not outright destruction, as they sailed past.

But this did not occur. Instead, Captain des Gouttes petitioned Governor Drucour on the very evening of Wolfe's landings that his fleet be permitted to weigh anchor and leave the harbor—not to engage the British, but to take the opportunity to escape to France. The governor convened a council of war and denied the request, reminding des Gouttes that the fleet had been sent to Louisbourg to bolster its defense and that such a flight would show a weakness to the enemy. "Being doomed to the same fate, it is proper to run the same risks," the governor told him.[26]

Drucour recognized the inevitable doom, but he was nonetheless determined to hold out as long as possible—not to win glory by a stand at Louisbourg, but rather to buy as much time as possible for Quebec. Every day that Drucour kept Amherst in the trenches in front of Louisbourg was one more day of summer that slipped away and would not be available to Amherst for an attack on Quebec. "It was a matter of deferring our fate as long as possible," wrote Drucour. "Thus, I said: if the French ships leave on 10 June . . . the admiral [Boscawen] will enter immediately after that. And in that case we would have been taken before the end of the month [June], and that would have given the attacking generals the advantage of using July and August to . . . take their ships up the river while the season was favorable to them."[27]

So, des Gouttes's ships remained at anchor in Louisbourg Harbor even as an advance guard of 400 colonial rangers and a main force of about 1,200 regulars led by Wolfe circled the harbor on the morning of June 12. Slipping past the French ships without notice, they quickly captured Lighthouse Point at the harbor's eastern mouth. Drucour had already abandoned the royal battery guarding the center of the harbor and had withdrawn almost all the troops into the main fortifications. So hasty was the French retreat from Lighthouse Point that Wolfe's men found French tents still standing, along with a large quantity of tools.

Only the small island at the mouth of the harbor remained in French hands to afford some measure of deterrence against Boscawen's ships sailing into the harbor and engaging the French men-of-war. Once again, des Gouttes and his captains appealed to Drucour that their ships be permitted to run the tightening gauntlet and make for the open ocean. Still Drucour declined. The governor avowed that the town was safe as long as the ships remained and that if they sailed and the town fell, he would charge des Gouttes and his squadron with its loss. On June 25, Wolfe's troops occupied the island battery, and the noose around Louisbourg Harbor was complete. Thereafter, the French sank four ships at the entrance to the harbor to block it and the majority of men, provisions, and powder were transferred from the remaining ships to the fortress.

Meanwhile, the British had erected a heavily entrenched encampment west of town on the same site used by Pepperell's colonial forces in 1745. Amherst seems to have taken his time with elaborate defensive preparations despite the fact that whatever thoughts Drucour might have had about venturing outside Louisbourg's walls to repulse the invaders vanished with St. Julien's rout from Wolfe's beachhead. By now, the British had landed heavy artillery and begun encircling the town and the full extent of the harbor in preparation for a full-scale bombardment.[28]

But such preparations did not stop the sort of chivalrous exchanges between attackers and besieged that seem to have been the norm of eighteenth-century warfare. Reports vary as to the exact transactions, but General Amherst appears to have started the exchanges when he dispatched a note to Governor Drucour and included with it a present of two pineapples for Madame Drucour. The governor responded with several bottles of champagne—the French, after all, had their traditions no matter how extreme the circumstances—which in turn were answered by more pineapples. This time, Madame Drucour responded with fresh butter, which she had made herself.[29]

There may, of course, have been ulterior motives in such exchanges, because they afforded each side the opportunity of a closer look at the other's fortifications while under a flag of truce. Even after Wolfe circled Louisbourg Harbor and captured Lighthouse Point, he "received a compliment, with a pyramid of sweetmeats, from the governor's lady, by a flag of truce sent to the eastern shore, by whom he returned his—with a pineapple, which he happened to have." Such European civilities caused one colonial observer to express great surprise at such "strange complaisance between inveterate enemies!"[30]

As June turned into July, Drucour had gotten the delay he had sought, but British artillery attacks were increasing and causing extensive damage to both the military and the civilian parts of the town. The French made one sortie outside the walls with about 1,000 men to disrupt the trenches closest to the Queen's Bastion, but then quickly retreated back inside the walls. On July 15, the frigate *Aréthuse* slipped out of the harbor and sailed for France with the news that only a miracle could now save Fortress Louisbourg.

Four days later, des Gouttes asked permission to burn what remained of his fleet, lest it fall into British hands. His ships had accomplished their purpose, as their firepower had added considerably to the cannon along Louisbourg's walls. But in turn, they had suffered great damage, notably the destruction of the sixty-four-gun *Célèbre*, the seventy-four-gun *Entreprenant*, and sixty-four-gun *Capricieux*, all of which were destroyed after a shell from one of Wolfe's heavy batteries exploded in the *Célèbre*'s powder magazine. The resulting fire quickly spread to the other two ships, closely anchored under Louisbourg's walls as they were, and all three burned to their water lines. Whether or not des Gouttes now had second thoughts about having failed to challenge Boscawen's landings has gone unrecorded.

The Dauphin Bastion had already sustained major damage, and the day after the three ships went up on flames, similar damage was visited upon the King's Bastion and the Citadel. The barracks in the Queen's Bastion were set afire the following night. Then, on the night of July 25, under the cover of yet another heavy fog, Admiral Boscawen dispatched the sloop *Hunter* and fireship *Aetna* into the harbor to seize the remaining two principal French ships. The British succeeded in towing off *Bienfaisant*, but *Prudent* stuck hard aground at low tide and was burned instead.

Now, whatever protections Drucour had sought from the French fleet were completely gone. Six of Boscawen's ships prepared to move into the harbor and bombard the town from its weakest side. Drucour summoned his officers early on the morning of July 26 and determined that it was finally time to ask Amherst for terms. An adjutant named Lippinot was sent to Amherst's headquarters. In one of the ironies of war, Lippinot had first rowed into Louisbourg from the *Tigre* back in 1749 to arrange the return of Louisbourg to the French after the Treaty of Aix-la-Chapelle. Now, nine years later, his journey was being reversed.

Drucour hoped for terms similar to those given the British garrison at Port Mahon on Minorca in 1756. Then, after capitulating, the British troops were merely transferred to Gibraltar. But Amherst and Admiral Boscawen—who had gone ashore to participate in the negotiations—were in no mood for such leniency. For one thing, the horrors of Fort William Henry were too fresh in their minds. There was also a growing frustration with the lateness of the season. Besides, where should the French be sent? Certainly not to Quebec. No, if Drucour wished to surrender, it would have to be unconditionally. For a moment, Drucour was again determined to fight, but then he was reminded that such a stance would only prolong the inevitable and cause grievous casualties among the town's civilian inhabitants. Reluctantly, he accepted the stringent terms.[31]

At eight o'clock the next morning, July 27, 1758, the Dauphin Gate of France's Atlantic fortress was opened and three companies of British grenadiers marched through it. "I have now the pleasure to write you," noted one British soldier, "that yesterday morning I was agreeably entertained with the 'Grenadiers March,' finely played, upon three fifes, and two drums relieving the grenadiers guard, under British colors, upon the walls of Louisbourg, which is a fine tune the French have not danced to for some time; but now it's time for 'em to pay the fifers." A few hours later, Brigadier Whitmore marched in with about 500 more men and took possession of the parade ground, where the French garrison was drawn up. The French officers were permitted to keep their swords, but that was the extent of Amherst and Boscawen's graciousness.

Louisbourg was in ruins from the British artillery fire. "The largest and best buildings in the city are reduced to ashes, and the rest all shattered and tore with shot and shells," continued the soldier.[32] All who had borne arms were to be sent to England as prisoners of war, including not only other troops on Cape Breton Island but also those on nearby Ile Saint Jean (Prince Edward Island). Those civilians who had not borne arms were to be transported to France at Admiral Boscawen's discretion.

According to another letter in the *Boston Gazette*, "The garrison seems as well pleased to leave Louisbourg, prisoners, as we conquerors, are to possess it."[33] Indeed, that was particularly true of a German regiment that had been enlisted by the French for service in Prussia, but had then been sent to Louisbourg four years before. By their own admission, these troops had been more slaves than soldiers and had "longed for a British invasion" to free them.[34]

There were others not so fortunate. There was to be no repe-

tition of Fort William Henry, but the massacre there the previous
year had by no means been forgotten. Revenge had already been
extracted by the rangers from Massachusetts who first came
ashore with Wolfe. Among the bodies of the French and their
Abenaki and Micmac allies that lay above Gabarus Bay were "one
hundred and odd French regulars and two Indians, which our
rangers scalped."[35]

Wolfe himself seems to have dismissed the Abenaki and Micmac with a sweeping generality of contempt. "These are a dastardly set of bloody rascals," he wrote. "We cut them to pieces
whenever we found them, in return for a thousand acts of cruelty
and barbarity."[36]

At least a few Frenchmen proved unhumbled by defeat. One
English observer noted that "the polite treatment which the
French have met with since the reduction of this place, has made
them extremely impudent," including an officer who struck the
coxswain of the barge ferrying him as a prisoner out to the waiting
British fleet. "And another French officer having the impudence
to run his hand under one of the Highlander's plaids, in an
improper place, the Highlander immediately eased him of his arm
and hand, by cutting them off with his broadsword."[37]

As Amherst pondered his next moves, he appointed Brigadier
Edward Whitmore temporary governor of Louisbourg. It was a bit-
tersweet assignment. Not only was Whitmore charged with governing a town largely in ruins, but Amherst's instructions gave him
the unpleasant task of deporting another generation of Acadians
from throughout Cape Breton and its neighboring islands. Little, if
any, thought was given to doing otherwise. Amherst was convinced
that their population was growing so rapidly that "many years
would not have passed before the inhabitants would have been sufficient to have defended it." Having just removed the French flag
from Louisbourg for the second time in a dozen years, the British
were determined that there not be a "next time." That the Acadians
might have been integrated into the British colony of Nova Scotia
and governed out of Halifax seemed to occur to no one.[38]

Perhaps an even greater irony was that a French priest in Acadia, witnessing the deportation of those who had tried for two generations to make a living on its rocky shores, could see more of the global picture than his sovereign had deigned to view. "What good is Louisbourg?" he asked. "It would be good if France were as strong at sea as England."[39] Indeed, when the front door of France's North American empire was closed, far more than Cape Breton Island reverberated with its slamming shut.

Meanwhile, in England, William Pitt stewed over the lack of news from North America. Pitt had gambled heavily on Amherst, and much if not all of his global strategy hinged on Amherst's success at Louisbourg. Returning to London after a few days of rest in the country, Pitt found Captain Amherst waiting for him and eagerly received the general's younger brother as "the most welcome messenger that had arrived in this kingdom for years."

After Captain Amherst laid out his report, the prime minister called it the greatest of news and hastened to advise Sir John Ligonier of his protégé's success. Both men went to report to the king, but with his typical Hanoverian stubbornness, George II seemed more concerned with reiterating his disapproval of Amherst's appointment in the first place than with celebrating his victory. In the face of this reaction, Pitt could afford to be smug. Things were finally beginning to go his way. Soon, all of England was celebrating the news.

"If I can go to Quebec, I will," Amherst had written in his dispatches, and Pitt saw no reason why he could not. Any day now might bring more good news from the other two prongs of Pitt's grand assault—Abercromby before Ticonderoga and Forbes en route to Fort Duquesne. Then, all that remained was for them to join Amherst in converging on Quebec. As always, Pitt's strategy was sound, but it would take a few more stumbles before it could be executed in full.[40]

8

"Till We Meet at Ticonderoga"

William Pitt's gamble in selecting Jeffery Amherst to command the Louisbourg campaign had paid off handsomely. Pitt was to be far less successful in his appointment of a commander to lead the center prong of his grand offensive. Lord Loudoun was out—there was no question about that—but if Pitt had acted precipitately in Loudoun's removal, as some would later contend, he compounded the error in his selection of a replacement. In truth, Pitt had little choice. George II had gotten his fill of Pitt's elevating junior officers and insisted that the top jobs of commander in chief in North America and commander of the thrust against Fort Carillon should both go to Loudoun's deputy, Major General James Abercromby.

Abercromby's redeeming feature—in George's eyes at least—appears to have been his longevity. At fifty-two, Abercromby had enjoyed a long, if unremarkable, military career. A Scot by birth, he had served as lieutenant colonel of the Royal Scots, had been a staff officer in Europe throughout the War of the Austrian Succession, had participated in expeditions against France, and in June 1756 arrived in New York as a member of Lord Loudoun's staff. Like Amherst, Abercromby had never held an independent command, although his preparation for one would at first glance seem to be equal to if not superior to Amherst's. The problem was that Abercromby was a plodder. Kindly, tactful, never robust in health, he was not one to inspire confidence in his subordinates or to rush into things. Without a doubt, he lacked the fighting spirit, quick wit, and decisiveness of Pitt's other new commanders.[1]

Pitt's orders recalling Loudoun and appointing Abercromby in his place reached New York on March 4, 1758. Among the accompanying dispatches were circulars to the various colonial governors. Loudoun had already asked them to provide "precisely 6,992 provincial soldiers" for the coming campaigns. Now, Pitt demanded that they raise upwards of 20,000. Fortunately, he also gave them important concessions over earlier calls for provincial troops. While he had yet to figure out how to pay for it, Pitt assured the colonial governors that—in the same manner as for British regulars—the crown would supply the required arms, ammunition, tents, and provisions for these new troops. He also declared—although this never came to pass—that he would make strong recommendations to Parliament to reimburse the colonies for their costs of raising, clothing, and paying these recruits. Of value to morale was the added concession that provincial officers as high as the rank of colonel would be accorded parity of rank with their counterparts in the regular army.

The response was astounding. On the morning after Pitt's letter was read to the Massachusetts assembly, the same legislators who had just refused to accord Lord Loudoun his request for 2,128 men voted unanimously to raise 7,000 on the terms proposed by Pitt. Other assemblies quickly followed suit. Even Connecticut, largely sheltered from the hazards of the front lines by New York and Massachusetts, voted to raise 5,000 troops. Within a month, Pitt's new policies had resulted in pledges to arm more than 23,000 provincials, plus thousands more to be employed as teamsters, bateaux operators, and craftsmen. Great Britain's North American colonies had finally been jumpstarted into a major war effort.[2]

These preparations and the prospect of another year's campaign soon produced a frenzy of activity and a high state of apprehension all along the New York frontier. Albany should have been used to such anxiety by now, as once again it became the focal point of preparations, but the situation was tense nonetheless. One evening a resident of Albany, whom the newspaper described as "being in liquor," failed to give the proper answer when challenged by a guard, and the poor inebriate was "fired upon by the sentinel, and killed on the spot."[3]

But at least that sentry was trying to do his duty. Some troops from Massachusetts on their way to Albany were camped west of Northhampton when boredom got the better of them. "A number for diversion, cut round the bottom of a large tree with their hatchets," reported the *Boston Gazette*. The tree fell on a tent occupied by four or five fellow soldiers. One was killed outright; another had an arm and both legs broken; and the others were badly bruised.[4]

Given such antics as these, what had the French to fear? First, of course, they had to fear the most penetrating of enemies. British intelligence from Fort Carillon confirmed what had now become the norm throughout Canada: "The French are much distressed for want of provisions, horseflesh runs very high in its price."[5]

For his part, Montcalm seems to have vacillated on this subject, one minute wringing his hands and saying that the state of provisions "makes me tremble," but the next minute, appearing quite unwilling to concede so much as an inch of ground. Despite dire shortages, Montcalm wrote to France in February 1758 that at a minimum any peace settlement should not only restore most of Acadia, but also permanently extinguish British claims to Lake Ontario, Lake Erie, and the Ohio Valley and limit British influence over the Six Nations.[6]

But what was Montcalm to do, militarily, in 1758 to accomplish this? Governor General Vaudreuil proposed that Montcalm and the majority of French regulars return to the Lake George country and make a powerful feint against Albany. Meanwhile, the chevalier de Lévis with a force of 3,000 handpicked regulars, Canadian provincials, and Indian allies would strike west, first to prevent the British from reestablishing Fort Oswego and then to spread terror down the Mohawk Valley. Vaudreuil hoped—with recent history as a guide—that the British would be paralyzed by the prospect of these two juggernauts and be left in confusion near Albany, not knowing which way to turn. If, in the resulting paralysis, Montcalm should be able to reduce Fort Edward with the same dispatch he had shown at Fort William Henry the year before, so much the better.

But besides the dismal lack of food and other provisions, there was one festering sore in this plan, indeed in the entire conduct of military affairs in New France. The relationship between Vaudreuil and Montcalm was slowly disintegrating. Vaudreuil, it will be remembered, had not taken kindly to Montcalm's arrival in the first place and each man had claimed the laurels after the French captured Oswego in 1756. At a time when New France required solidarity, "there existed a bitter rivalry, a personal animosity, between its highest executive and its most gifted general that boded no good for the future." A superior in France counseled Montcalm that now was "not the moment to appraise the qualities and talents of those who direct affairs," but the growing feud would simmer throughout 1758 before erupting the following year with disastrous results.[7]

"We are on the eve of the most cruel famine," wrote a resident of Quebec on May 19, 1758, after the daily ration of flour had been cut in half to two ounces per person. About 1,200 to 1,500 horses had already been purchased to "make up for the want of bread, beef, and other necessaries of life." Only arrival, in the nick of time, of five flour-laden ships from Bordeaux, which had somehow managed to elude the British fleet, gave the colony some measure of sustenance and permitted the dispatch of troops into the field.[8]

As spring turned to summer, Montcalm hurried south to reinforce Fort Carillon at the place the Mohawk called Ticonderoga, "the place between the great waters." A short time later, reports that the British were massing north of Albany in unprecedented strength caused Governor Vaudreuil to send Lévis in support of

Montcalm rather than on his western strike. Time would tell whether it was the correct decision.

The British forces were indeed assembling north of Albany in record numbers, but there were many in the cadre of British officers who held a decidedly low opinion of this new influx of provincial recruits. Even after receiving their assistance at Louisbourg, Brigadier James Wolfe was among those voicing the most disdain. "The Americans are in general the dirtiest most contemptible cowardly dogs that you can conceive," wrote Wolfe. "There is no depending on them in action. They fall down in their own dirt and desert by battalions, officers and all. Such rascals as those are rather an encumbrance than any real strength to an army."[9]

There was, however, to be at least one colonial who would elevate himself and a group of like-minded woodsmen to a level of esteem among certain British officers. His name, of course, was Robert Rogers, and he and his rangers would come to be called the original Green Berets of the American army. Curiously, the major histories of this era give little insight into the man. Perhaps it is assumed that Rogers is so well known that the writer need do no more than breathe his name. More likely, these writers, too, have struggled to cut away 250 years of legend and expose the facts.[10]

Robert Rogers was born in the tiny settlement of Methuen on the northern frontier of Massachusetts Bay Colony on November 18, 1731. His parents were Scots from Northern Ireland who appear to have immigrated to Massachusetts "only a short time before Robert's birth." As Presbyterians, they were unwelcome in many New England towns, but the outskirts of Methuen were home to a number of Ulster Scots who had settled there on common ground. It wasn't long, however, before the promise of individual ownership of new land led his family into the Merrimack River valley of New Hampshire. Here, young Robert would hone his skills as a woodsman and have his first taste of Indian warfare.

In 1745, during King George's War, the French encouraged

Indian attacks all along the New England frontier. Robert Rogers was fourteen that fall and, "by frontier standards, old enough to take his place among men." By the time the conflict ended four years later, Rogers was a seasoned warrior and, along with a fellow New Hampshire lad named John Stark, eager to be among those to explore farther west into the Connecticut River valley. Such advances by settlers met with renewed resistance from the Indians, particularly by the Abenaki of the village of Saint Francis, and Rogers served in a variety of militia assignments.

Whatever his growing reputation as a "ranger," Robert Rogers also had his first brush with the dark side that would plague him all of his life. Short on cash and unable to make payments on land he had purchased in Merrimack, he fell under the spell of a counterfeiter named Owen Sullivan. Although not directly engaged in the counterfeiting, Rogers appears to have been among those who knowingly passed false bills obtained from Sullivan. Rogers was arrested along with eighteen other suspects, but never tried. This was in part because the principal culprit, Sullivan, had managed to slip away, but also because now, in the spring of 1755, New Hampshire was suddenly occupied with a far more pressing concern: another war with France. Rogers once more volunteered for duty and promised to recruit a company of men. On April 24, 1755, when the New Hampshire regiment was officially activated, twenty-three-year-old Robert Rogers became captain of "Company One" with John Stark as his lieutenant.[11]

Numerous assignments followed. Rogers and his men convoyed supply trains north from Albany, scouted French positions along Lake George and Lake Champlain, and captured prisoners needed for interrogations. These assignments were frequently far afield and caused him to miss the two major engagements of those early years: Dieskau's attack against the British at Lake George in 1755 and Montcalm's siege of Fort William Henry in 1757—during which Rogers was embarked on Lord Loudoun's stillborn assault against Louisbourg. But Rogers and his "rangers," as they came to be called, still saw plenty of action.

In January 1757, Rogers led eighty-five rangers north on ice skates across frozen Lake George and then on snowshoes over the mountains to halfway between forts Saint Frédéric and Carillon. Their mission was to gather intelligence and disrupt the lines of communication and supply between the two posts. The rangers captured several sleighs and seven prisoners, but then learned that two hundred Canadians and forty-five Indians had just arrived at Carillon and were poised to take to the woods and cut off their avenue of retreat.

A fierce fight broke out in the wintry woods. For a time, Rogers and his men appeared to be hopelessly surrounded; and the Canadians, singling out Rogers by name, repeatedly called for his surrender, all the while praising the valor of his men. Although Rogers was wounded in both the head and the wrist, he later reported that he and his men were "neither to be dismayed by their threats, nor flattered by their professions, and determined to conquer, or die with arms in their hands." That night, under the cover of darkness, the rangers followed one of their cardinal rules of warfare. They broke off their engagement with a superior force, divided into small groups, and melted away into the forest to rendezvous again at a predetermined location.[12]

Such exploits, while hardly stunning military victories, nonetheless made quite a name for Robert Rogers and his men. In fact, colonial newspapers from this period are filled with reports of their adventures. "We hear that the famous Rogers of Fort William Henry has made another tour near the French encampment at Crown Point [Saint Frédéric]," reported the *New York Mercury* as early as May 1756. By August 1757, a paper in Boston was calling him "Major Rogers." A lengthy account by Rogers himself of another scouting expedition against Fort Carillon in force—"the famous Rogers," he was again called—appeared soon afterward.[13] Had this been the twentieth century, one might have been inclined to look around for his press agent.

But there was always the dark side, too. Occasional defeats in the field were to be expected, but lack of discipline by his rangers and even talk of that ugliest of military words—mutiny—were quite another. During the winter of 1757–1758 considerable friction developed between British regulars stationed at Fort Edward and the rangers of Rogers's command camped below the fort on Rogers Island in the Hudson River. In November, the rangers' discipline was less than its best on an unsuccessful raid that Rogers missed because of scurvy. (Shooting at game as one passed and ignoring guard duty were definitely not condoned by the rangers' rules.) This led General Abercromby to report that if the ranger companies were to be increased as Rogers had requested, "it will be necessary to put some regular officers amongst them to introduce a good deal of subordination."

The symbol of such subordination that came to grate on the rangers was the six-foot whipping post embedded in the ground on the island. It was "almost the symbol of discipline in the regular army" but was met with great disdain by most colonials. When two rangers were whipped for stealing rum from British stores, word quickly spread that "if rangers were to be flogged, there would not be rangers." In the fervor that followed—no doubt caused partly by the boredom of a cold winter camp—a group of grumbling rangers surrounded the whipping post and one of them cut it down with a few well-placed blows of an ax.

An uproar ensued on the island and was heard by Fort Edward's commandant, Colonel William Haviland. Before the matter was put to rest, Rogers stood before Haviland pleading for leniency for those involved and warning of mass desertions if it was not granted. For his part, Haviland proposed that "it would be better they were all gone than have such a riotous sort of people," but if Rogers would catch anyone who attempted to desert, Haviland would "have him hanged as an example."

Haviland passed the proceedings of Rogers's own inquiry into the disturbance on to Abercromby and recommended various court-martial charges "to prevent these mutinous fellows escaping a punishment suitable to their crime." Meanwhile, Rogers did the only thing he could think of to blunt such criticism: he promptly marched off with 150 men to make yet another raid against the French at Fort Carillon. By the time they returned, the entire sordid affair seems to have blown over, largely because even the staid Abercromby recognized the value of Rogers and his rangers. Whereas most British officers were content to hunker down behind log walls for the winter, Rogers and his men were among the few determined to carry the fight to the enemy no matter what the season.[14]

Doubtless it came as some surprise, however, that Robert Rogers was to find his greatest champion in the form of an affable and competent English lord. If Pitt was indeed stuck with James Abercromby as his commander in chief in North America, he was absolutely determined that a far more able and assertive field commander be at his side to direct the assault against Fort Carillon. Pitt found that man in George Augustus, viscount Howe, and appointed him to be Abercromby's deputy.

Lord Howe, born in 1725, was the eldest of three brothers destined to distinguish themselves in Great Britain's service. He chose the army, and like so many of his generation, saw service in Flanders and elsewhere on the continent under the duke of Cumberland. He embarked for America to serve as Abercromby's deputy with both a sense of adventure and curiosity.

The second son, Richard, chose the sea and saw duty in both the merchant marine and the Royal Navy before being commissioned a lieutenant in 1745. Assignments of increasing responsibility followed until he was given command of the sixty-gun frigate *Dunkirk* in January 1755. It was as captain of the *Dunkirk* that he captured the French frigate *Alcide* off Cape Race, Newfoundland, the following June. Richard Howe would go on to become a vice admiral and commander in chief of British naval forces in North America during the opening years of the American Revolution.

The third son, William, began service in the duke of Cumberland's light dragoons in 1746, was appointed major in the Sixtieth Foot Regiment in 1756, and most recently had been engaged under Wolfe at Louisbourg. William Howe's name would reverberate in North America for the next twenty years, from the heights of Quebec with Wolfe to the forests of the Brandywine against George Washington.

Lord Howe was everything that the stuffy and ineffective Abercromby was not, and British regulars and colonials alike soon came to revere him. "Lord Howe was the idol of the army," a provincial carpenter wrote, "in him they placed the utmost confidence. From the few days I had to observe his manner of conducting, it was not extravagant to suppose that every soldier in the army had a personal attachment to him. He frequently came among the carpenters, and his manner was so easy and familiar, that you lost all that constraint or diffidence we feel when addressed by our superiors, whose manners are forbidding."[15]

Robert Rogers was commissioned "Major of the Rangers in His Majesty's service, and captain of a company of the same" on April 6, 1758, and ordered to report to Lord Howe at Albany. Rogers did so and "had a long conversation with him upon the different modes of distressing the enemy, and prosecuting the war with vigor."[16] Indeed, Howe was so impressed with Rogers's tactics that he went on at least one scouting foray with the rangers in order to learn more. Far from indoctrinating the rangers into the regimen of European warfare as Colonel Haviland had recommended, Howe saw to it that the rangers' style was impressed on the other units of Abercromby's command.

"You would laugh to see the droll figure we all cut," read one report from north of Albany in early June; "regulars and provincials are all ordered to cut the brim of their hats off. Even the General himself is allowed to carry no more than a common private's tent. The regulars as well as provincials have left off their proper regimentals, that is, they have cut their coats so as scarcely to reach their waist:—You would not distinguish us from common plowmen." No women were allowed to follow the camp, and it was further noted that in the matter of washing clothes, "Lord Howe . . . has already showed an example, by going himself to the brook, and washing his own linen."[17]

Just how completely Lord Howe embraced this wilderness life is best illustrated by an often-told story of a dinner he gave for his officers just before the army embarked on Lake George. Entertaining one's junior officers in style was an accepted requirement of command in those days, and Howe's subordinates no doubt arrived at his tent relishing the prospect of a brief respite of English grace and decorum. Instead, they found rough logs in place of chairs and bearskins thrown down on the ground instead of a carpet. There was not a piece of china or silver to be seen. In due course, a servant set a blackened pot of pork and peas on one of the bearskins.

With a knowing glance about him, Howe took a sheath containing a knife and fork from his pocket and began to cut away while his guests looked on in astonishment. "Is it possible, gentlemen," his lordship asked, "that you have come on this campaign without providing yourselves with what is necessary?" Polite laughter turned to sighs of relief only as Lord Howe moved to distribute a similar knife and fork set to each of them. It was a far cry from what might have been expected in similar circumstances in Europe; but this was not England, and Lord Howe seems to have been the most astute of the British officers in recognizing that difference. For their part, the colonials idolized him for doing so.[18]

By June 8, 1758, Lord Howe was encamped at Fort Edward with about half of Abercromby's army. Two weeks later found them at the ruins of Fort William Henry on the southern end of Lake George. Upwards of 500 wagons were at work between Albany and the lake hauling supplies and matériel northward after them. On Lake George itself, more than 1,000 bateaux, "each capable of carrying twenty-five men," were being readied for the assault down the lake.

Contemporary accounts speak of building boats to "cross the lake," but to one who has not wandered along Lake George's forested shores, those words do not paint the full picture. To be sure, the object was to "cross the lake," but Lake George is one to three miles wide and thirty miles long. What was now required was to row from the ruins of Fort William Henry north down the lake to its northern outlet, which led through a short passage before emptying into Lake Champlain within an easy artillery shot of Fort Carillon.[19]

Major General Abercromby arrived at the Fort William Henry camp with the remainder of his army on June 28, but it was still very much Lord Howe's show. Howe sent Rogers north to make a detailed reconnaissance of Fort Carillon and the line of advance, including the best landing site at the lake's northern end and the approach to the fort. He also ordered Rogers to survey Lake Champlain for three miles beyond Carillon and "discover the enemy's forces in that quarter." Howe, it seems, was being quite thorough.[20]

Reports of the exact numbers of troops engaged in the campaign vary, but most speak of about 15,000—approximately 6,000 British regulars and 9,000 colonials, easily the largest army yet assembled on the North American continent. Thirty-five miles away at Fort Carillon, Montcalm was doing his best to scrape together an opposing force of less than 4,000. "In all human probability," wrote one correspondent, "a few days will decide the dispute."[21]

Early on the morning of July 5, 1758, Abercromby's army—under Howe's able command in the field—took to its boats and sailed north down Lake George. What a sight that must have been! The glint of thousands of muskets and the flutter of regimental colors filled the lake as the dull clunking of oars and the staccato beat of drums resounded from the hillsides. By one account some 900 bateaux and 135 whaleboats were laden with men and accompanying baggage, stores, and ammunition. Artillery pieces followed on heavy flatboats. One wounded officer wrote from Albany two weeks later, "I never beheld so delightful a prospect."[22]

Down the lake they went, first to Sabbath Day Point, twenty-five miles away, by late afternoon. After briefly resting and allowing the long line of boats to close up, the advance guard started off again just before midnight. It was led by Rogers and Lord Howe in the first boat, but also in the vanguard was a light infantry regiment commanded by Lieutenant Colonel Thomas Gage, the same officer who had led Braddock's advance across the Monongahela three years before. Designed to be more mobile and more rapidly deployed than regular British regiments, this light infantry regiment was the British army's way of having regular troops that could imitate the tactics of Rogers' Rangers without some of the rangers' antics. It would prove to be the beginning of light infantry units in the British army.[23]

Dawn found the leading boats passing under the granite slabs of Rogers Rock, which rises above the western bank of the lake near the narrows at its northern end. (Legend subsequently suggested that earlier the same year Rogers had escaped from a pursuing party of French and Indians by sliding down this precipice—on his seat or on snowshoes; take your pick—but Rogers himself fails to mention it in his journal.[24]) On this July morning, Rogers was no doubt much more concerned about a French advance party of some 350 regulars and Canadian provincials who watched every movement of the British from the heights.

By noon on July 6, the bulk of the British forces were ashore at the northern end of the lake. Hindsight suggests that Howe should have ordered the entire army to follow Rogers along the route he had previously reconnoitered and close with Fort Carillon as quickly as possible. But this was easier said than done. Montcalm had perhaps as many as 1,500 troops spread out along the connecting stream between Lake George and Lake Champlain. There was also the French advance guard, now descending Rogers Rock and trying to make its way back to the main French lines. Finally, there was the forest itself. A dense thicket of timber and undergrowth extending down to the water's edge gave grim testimony as to why the British had opted to sail the thirty miles down Lake George in the first place.

According to Rogers, it was General Abercromby who directed him to take some rangers and seize high ground about a mile inland from the landing place. This left Lord Howe to advance with other ranger units and elements of Gage's light infantry. In the tangled woods directions quickly became confused, and any sense of a line of battle disintegrated. What does appear to have happened, however, is that Lord Howe remained on the leading edge of the confusion, gallantly indeed, but this was hardly the place for the man Wolfe later called "the spirit of that army, and the very best officer in the King's service."[25]

Howe's advance guard collided in the deep woods with elements of the French advance guard rushing north from Rogers Rock to avoid being cut off. "A sharp fire commenced in the rear of Lyman's [Connecticut] regiment," Rogers recorded, and when it was over, the French had been routed but Lord Howe was dead, killed almost instantly by a musket ball through his chest. One unsubstantiated report suggested that Howe had been calling for the surrender of a French officer when he was struck, or he simply may have been hit in a hail of blindly fired bullets. The result was the same. "So far things had been properly conducted and with spirit," wrote Captain Joshua Loring of the Royal Navy, "but no sooner was his lordship dead, than everything took a different turn and finally ended in confusion and disgrace."[26]

With the death of Lord Howe, both the inertia and the intelligence of the British advance faltered. The advance guard spent the remainder of the day disentangling itself from the wooded maze and then returned to the landing site to regroup. Abercromby was two hours' march from Fort Carillon, but he would spend two days getting there. Yes, the woods were dense; yes, there was the prospect of an attack by the French; but rather than strike quickly, Abercromby dallied and constructed an entrenched camp at the site of a French sawmill just two miles from the fort. "I can't but observe," recorded a physician from Massachusetts named Caleb Rea, "since Lord Howe's death business seems a little stagnant."[27]

Abercromby's delay was Montcalm's good fortune, and Montcalm made the most of it. He had arrived at Carillon only the week before to find a dismal shortage of both men and supplies—by

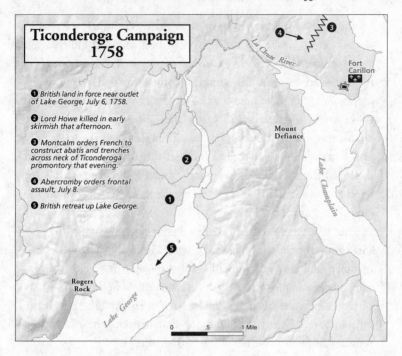

Ticonderoga Campaign 1758

❶ British land in force near outlet of Lake George, July 6, 1758.

❷ Lord Howe killed in early skirmish that afternoon.

❸ Montcalm orders French to construct abatis and trenches across neck of Ticonderoga promontory that evening.

❹ Abercromby orders frontal assault, July 8.

❺ British retreat up Lake George.

La Chute River

Fort Carillon

Mount Defiance

Lake Champlain

Rogers Rock

Lake George

0 .5 1 Mile

now the modus operandi of all military operations in New France. As the British force passed under Rogers Rock, Montcalm pondered his fate. Where should he make his stand? There was little chance of holding Fort Carillon against a protracted siege. For one thing, he and his men would soon starve to death. If things went poorly, Montcalm stewed, Fort Carillon might turn out to be Fort William Henry in reverse.

Several of Montcalm's lieutenants favored making a stand at the sawmill and in the woods along the outflow stream from Lake George. Others warned that this location was susceptible to enemy fire from the surrounding heights, parts of which Rogers was indeed soon sent to seize. So, on the evening of July 6, as Abercromby's army regrouped from the skirmish that had cost it Lord Howe, Montcalm made his decision. He ordered his troops to undertake a flurry of construction across the narrow neck of the Ticonderoga promontory northwest of the fort.

From the fort at the eastern tip of the promontory, the terrain sloped gently westward and then rose to a rounded hill that fell away toward the most likely route of the British advance. Just below the crest of this hill, Montcalm's men dug shallow entrenchments and then cut huge logs to stack above them. In front of these works for nearly 100 yards, they piled a series of deadly abatis, tree limbs strewn about like a tangled web. Leaving only one battalion to man the artillery in the fort, Montcalm then deployed the majority of his troops, including recent arrivals led by the chevalier de Lévis, along this line. Providence help the enemy that advanced on them.

But truth be told, their present enemy had no compelling reason to advance directly against them. Montcalm prepared for a frontal assault, but left his flanks sadly exposed. To be sure, the terrain there was much more rugged, but overwhelming numbers thrown against either flank might have been enough to turn the outnumbered French line. Or, for that matter, why not simply reduce this formidable barricade with artillery?

To the southwest of Fort Carillon and just over a mile away, the rounded hump of Rattlesnake Hill (later renamed Mount Defiance) rose some 700 feet above the waters of Lake Champlain and the outflow stream from Lake George. Montcalm's haste and lack of manpower had by necessity left its slopes undefended. Having taken great pains to float an artillery train of sixteen cannons, eleven mortars, thirteen howitzers, and 8,000 rounds of ammunition down Lake George, Abercromby might now have taken what time he needed to array his cannons on Rattlesnake Hill and blast the French out of their entrenchments. But having dithered when he should have charged, Abercromby now charged when he should have waited for his artillery to be dragged into place. Nineteen years later, during the American Revolution, the advancing British recalled the lesson and in a day's time placed cannons on the slopes of Mount Defiance. The Americans surrendered the fort without firing a shot.

• • •

Early on the morning of July 8, Abercromby sent a junior officer forward to make a reconnaissance of the French lines. After only a cursory look, he returned with the assertion that the trenches could be carried by a frontal assault. Abercromby did not ask for a second opinion from the more experienced officers, including Thomas Gage, who with Howe's death was suddenly second in command. Indeed, the only question that Abercromby asked his staff, before ordering them into an eighteenth-century forerunner of the charge of the Light Brigade, was whether they wished to advance in ranks of three deep or four deep. Three deep was the answer, and while his artillery remained at its landing site, Abercromby's army deployed in long ranks across the front of the French line. Perhaps for a moment, even Montcalm could not believe his good fortune.

The advance guard of Rogers's ranger companies, Gage's light infantry regiment, and a battalion of Massachusetts light infantry moved forward to the edge of the abatis tangle, chased off a few French pickets, and set to sniping at any heads that poked up above the crest of the main fortifications. This, as the historian Fred Anderson points out, was at least some evidence that Gage, if not Abercromby, had learned a lesson or two from his days on the Monongahela. But the array of ranks spread out three deep behind them belied that education and showed that Abercromby "intended to use his most thoroughly disciplined troops in the most thoroughly conventional way: by arraying them in three long, parallel lines and sending them straight up against the French barricade."[28]

It was noon by the time eight regular regiments, backed by six provincial regiments in reserve, dressed their ranks and prepared to advance into the hornet's nest. Then, to the beat of drums and swirl of bagpipes—the Highlanders Black Watch regiment was in the center of the line—the thin line started across the field.

The Black Watch was another of Pitt's controversial military experiments. The Scots Highlanders had long proved that they were fighters, but now that they were finally a lasting part of the British Empire, Pitt determined that they should use their bravery fighting for it, not against it. To the generation of English military leaders who had fought against certain Scots at Culloden only a decade before, it was a shocking proposition. A Highlander regiment had acquitted itself well with Wolfe at Louisbourg, and now the Black Watch was about to prove its mettle before Carillon.

Onward they came. What discipline, what courage, what waste! Regiment after regiment approached the edge of the abatis and disintegrated, torn from its formation by the tangle of fallen timber and struck down by the hail of bullets from the fortifications. March up briskly, rush the enemy's fire, and pour volleys into the trenches behind the breastworks, Abercromby had ordered. But the advancing regiments were chewed up long before they reached it. "Our orders were to [run] to the breastwork and get in if we could," a survivor recalled, "but their lines were full, and they killed our men so fast, that we could not gain it."[29]

And what of the Black Watch, the Highlander regiment that some had questioned as "Pitt's grand experiment"? In the center of the British line, the Black Watch had taken "as hot a fire for about three hours as possibly could be," all the time seeing nothing of the French "but their hats and the ends of their muskets." Casualties were appalling. By one count, out of approximately 1,000 men, the regiment lost 8 officers, 9 sergeants, and 297 soldiers killed and 17 officers, 10 sergeants, and 306 soldiers wounded—a horrific casualty rate of 65 percent. Among those wounded was fifty-five-year-old Major Duncan Campbell.[30]

Those are the facts. Here is the legend. After the last gasp of the Stuarts at Culloden, a man with torn clothing and a kilt covered with blood appeared at the gate of the venerable castle of Inverawe on the banks of the River Awe in the western Highlands. The laird of the castle who answered the summons was Duncan Campbell. The stranger confessed to having killed a man in a brawl and pleaded for sanctuary from his pursuers. Campbell vowed that he would provide it and even acknowledged the stranger's plea that he so "swear on your dirk."

Scarcely had the stranger been safely hidden then there came another pounding at Inverawe's gate. Two armed men confronted Duncan Campbell with the news that his cousin, Donald, had been murdered and that the killer was fleeing before them. Much distressed, Duncan denied any knowledge of the unwanted guest he had sworn on his dirk to protect and sent the men away.

That night sleep came hard to the laird, but when it finally did come, he was confronted by the ghost of his murdered cousin standing beside his bed. The figure implored Inverawe to "shield not the murderer." In the morning, Duncan went to the stranger and asked him to leave. "But you have sworn on your dirk," came the stranger's reply. Now, even more distressed, Campbell led him out of the castle and hid him in a nearby cave. That night the same ghostly presence returned to his bedside and once more implored him to "shield not the murderer." In the morning, Campbell hastened to the cave to oust his unwelcome guest, but the murderer was gone. Once again that evening, sleep came fitfully, but the ghostly vision of his departed cousin returned a final time. "Farewell, Inverawe," was all that he said, "Farewell, till we meet at Ticonderoga."

The strange name, so the story goes, dwelled on Duncan Campbell's mind until a few years later, when, as a major in the Black Watch, he found himself rowing toward Ticonderoga. On the night before the attack on the abatis, the ghostly vision of his cousin appeared to him again. In the morning, despite well-meaning comrades who tried to dissuade him from his gloom, Campbell averred, "This is Ticonderoga. I shall die today!"

Major Duncan Campbell survived the day, but died of his wounds nine days later in the hospital at Fort Edward. He now lies buried in the Union Cemetery in Fort Edward. The legend may be denied, but what cannot be denied is that Abercromby's misuse of these brave and loyal troops was enough to give any Scot nightmares.[31]

Back at the sawmill camp, Abercromby had heard the terrible din of battle, but had not dared to venture forward to witness it.

Lord Howe had lost his life by being willing to lead his army. Abercromby almost lost his army by refusing to lead it. By late afternoon there was a panicked retreat back to the landing site on Lake George. Upwards of 2,000 troops had fallen dead or wounded, and rumors abounded that the French were counterattacking and about to drive the rest into the lake.

At some point it must have dawned on even the blundering Abercromby that he had not only lost a large portion of his army but opened a gaping hole in the New York frontier. Far from taking any blame himself, however, he simply reported to Pitt that "it was therefore judged necessary for the preservation of the remainder of so many brave men, and not to run the risk of the enemy's penetrating into His Majesty's dominions, which might have been the case if a total defeat had ensued, that we should make the best retreat possible."[32]

After such a misinformed and imprudent attack with such resulting death, it is hard to speculate what "a total defeat" might have been. Official returns later put the total number of British casualties at 1,944—of which 551 were killed; 1,356 were wounded; and 37 were missing. The majority of the casualties were regulars. French losses were 106 killed and 266 wounded.[33]

Dawn on the morning of July 9, 1758, found the largest British army yet assembled in North America rowing with all its might back up Lake George. The images of fluttered banners and glistening bayonets had been replaced by chaos. Ironically, the British were fleeing an army that was but a quarter of the size of their own and that was not in pursuit. Montcalm was more than content to maintain his position at "the place between the waters" and declare victory.

In fact, Montcalm was ecstatic, and his ecstasy caused him to misstate the facts in a letter to his wife. He had not stopped an army of 25,000 and had not inflicted 5,000 casualties, as he wrote to her; but just the same he had saved New France for another summer. Later Montcalm wrote to a friend: "The army, the too-small army of the King, has beaten the enemy. What a day for

France! If I had had two hundred Indians to send out at the head of a thousand picked men under the Chevalier de Lévis, not many would have escaped. Ah, my dear Doreil, what soldiers are ours! I never saw the like. Why were they not at Louisbourg?"[34]

Why indeed, but the far more pressing question was why Abercromby had rushed headlong into defeat with an abandon that even Braddock had not evidenced on the Monongahela. Two contemporary poems emerged from the carnage to offer their authors' own explanations. One bemoaned the loss of Lord Howe:

> *Illustrious Man! Why so intrepid brave.*
> *Since thine was like to prove the army's grave?*
> *Could not that love for us, that led thee on,*
> *Show, that by losing thee, we were undone?*
> *Why was that life in which our hopes were centered,*
> *So much exposed, and too, too much adventured?*
> *Of arms and arts and all that's good a mirror;*
> *Thy too much braveness was thy only error,*
> *Now rest in joy a peace and let thy name*
> *Ascend illustrious in the sphere of fame.*[35]

The other poem was written by Montcalm himself and inscribed on a great cross that he ordered erected on the battle-field the morning after the battle:

> *To whom belongs this victory?*
> *Commander? Soldier? Abatis?*
> *Behold God's sign! For only He*
> *Himself hath triumphed here.*[36]

9

THE BATEAU MAN

Basking in the triumph of his victory at Louisbourg, Major General Jeffery Amherst considered his next move. Thanks to the stubborn if somewhat reluctant defense of Louisbourg by the French navy, half of the summer of 1758 had passed before the town's capitulation. The days were growing shorter. Amherst's redheaded, hot-tempered young brigadier, James Wolfe, advocated pushing on to Quebec no matter what the season, but Amherst was uneasy about how his commander in chief, Major General James Abercromby, was faring north of Albany.

Then on August 1 "came the news which Amherst had feared, only ten times worse. Abercromby had been not only defeated at Ticonderoga, he had been slaughtered." With their Lake Champlain front secure, it now seemed that the French could concentrate their forces against any army that Amherst might send up the Saint Lawrence, or they could advance on Albany at their pleasure. Amherst and Admiral Boscawen held a hurried conference and reluctantly wrote a joint report to Pitt detailing the impracticality of attacking Quebec that summer. Rather than sail for Quebec, Amherst would rush to Abercromby's aid, while Wolfe was given the rather unpleasant task of destroying generally unarmed French fishing villages on the Gulf of Saint Lawrence.[1]

* * *

Meanwhile, on the southern shores of Lake George, Abercromby took stock of his own situation. What Wolfe characterized as "the unlucky accident that has betaken the troops under Mr. Abercromby," was much more than that. Although Montcalm did not mount a major offensive southward in pursuit, he did send about 1,000 men around the eastern side of Lake George to disrupt Abercromby's lines of communication between the lake and Fort Edward. The French fell on a wagon convoy and killed 116 men, sixteen of whom were rangers. Rogers was dispatched with a party of about 700 men, including rangers, Connecticut provincials under the command of Major Israel Putnam, and some of Gage's light infantry, to cut off their retreat.

The main French force eluded him, but as Rogers returned to Fort Edward, other French troops attacked and Major Putnam was among those taken prisoner. After about an hour's fighting, Rogers nonetheless claimed that "we kept the field and buried our dead." All this left Abercromby feeling quite vulnerable, particularly because he had finally given Colonel John Bradstreet permission to undertake a daring plan that Bradstreet had been promoting for at least three years.[2]

By most accounts John Bradstreet was a doer, a man of great energy and personal courage who had worked his way up the hard way. He seems to have been driven by an insatiable ambition that occasionally found him crossing the line into dubious activities. "John Bradstreet," wrote the historian Harrison Bird more succinctly, "was a bateau man."[3]

Indeed, no one had mastered the use of those ugly flat-bottomed boats for operations or logistics more effectively than John Bradstreet. The wooden boats certainly weren't pretty to look at or elegant under way—propelled by oars, makeshift sails, or both—but in this war in this roadless wilderness, bateaux were the counterparts of modern half-tracks or Bradley fighting vehicles. One bateau was capable of transporting twenty-five men and supplies along watery routes for great distances.

John Bradstreet was born in Nova Scotia in 1714, the son of a British army lieutenant and an Acadian mother. From the age of fourteen onward, he took the British army as his avenue of promotion, and he steadfastly stuck with it despite agonizingly slow advancements. At a chance meeting with Governor William Shirley of Massachusetts in 1744, Bradstreet impressed Shirley with his intimate knowledge of Nova Scotia and may have persuaded Shirley that an attack on Louisbourg could be successful. When Shirley sent William Pepperell's expedition in that direction the following year, Bradstreet was in the thick of things. Bradstreet remained in the British army after the peace of Aix-la-Chapelle and seems to have relied on his mother's Acadian roots to carry on questionable trading practices—if not outright smuggling—with the French while he was a British officer.[4]

By 1755, Bradstreet was a captain, and from Oswego he reported to Governor Shirley—by then commander in chief in North America—that the French were marshaling their forces at Fort Frontenac on Lake Ontario and were likely to attack Oswego. "Frontenac" was a name that Bradstreet would come to repeat time and time again.[5]

The following year, observing that operations at both Lake George and Oswego had suffered in 1755 "for want of a sufficient number of wagons, horses, and bateaux-men," Shirley informed his superiors in London that he had taken steps to prevent the same from happening again. "I gave orders for engaging 2,000 bateau-men to be disposed into companies of 50 men each under the command of one captain and an assistant, and to be put under the general direction of one officer well skilled in the many branches of this important trust."

In other words, Shirley had found someone who not only knew about boats and logistics but also knew how to fight. That man proved to be John Bradstreet. Shirley promoted him to the rank of lieutenant colonel and put him in charge of bateau service on the Mohawk River, essentially the supply line between Albany and New York's western outposts, including Oswego.[6]

If nothing else, this proves that Shirley—despite subsequent criticism—understood that a wilderness war could never be won without reliable lines of supply. These tasks had to be undertaken not by unarmed teamsters and boatmen, but by men able to fight their way through no matter what the terrain or the opposition. When Oswego fell in the summer of 1756, Shirley bore the brunt of the blame, but not because of any problems with supply: six weeks earlier Bradstreet's bateaux convoy had "thrown into Oswego six months provisions for five thousand men . . . and defeated a party of French and Indians on his way back."[7]

Throughout these years of lackluster British military achievement in New York, Bradstreet performed admirably and sang but one song: Frontenac. Never mind Britain's loss of Oswego, or Webb's failure to reinforce Fort William Henry the following year. Bradstreet wanted to lead a daring raid against the hub of New France's own western supply line—Fort Frontenac. No doubt it was at Bradstreet's urging that first Shirley and then Lord Loudoun were inclined to strike this way, but the exigencies of the moment always prevented them from turning Bradstreet loose.

Never shy about stating his own qualifications or promoting his own advancement, Bradstreet continued to beat the drum not only for an assault on Frontenac, but also for an attack from there down the Saint Lawrence to Montreal. Mincing no words, Bradstreet unabashedly proclaimed: "Were it not owned by all degrees of people that no person in America is more capable of conducting an inland expedition in these parts than I am, I would by no means be so presumptuous as to offer myself as a candidate for the command."

Bradstreet continued, "I know and am sensible of the distance, difficulties, and dangers which would attend it, and believe few there are who think it practicable, but . . . so far am I convinced of my being able to go through with it that I will risk my reputation upon the success of it as far as the taking of Montreal and all the forts upon the lakes and join the fleet should they reach Quebec—unavoidable accidents excepted."[8]

This was bold talk, particularly as the war in North America seemed to be filled with "unavoidable accidents," but Bradstreet was continually diverted to other missions. Most recently, Abercromby had summoned him to Ticonderoga. Bradstreet had been in the vanguard there with Rogers and Howe, and his bateau men had done yeomen's service marshaling Abercromby's army forward and shepherding its retreat. Now, after the staggering setback, Abercromby "was looking desperately for something that might soften the bitterness of his defeat."

Once more, Bradstreet pleaded his case for an attack on Fort Frontenac. On July 13, 1758, Abercromby agreed. Bradstreet was directed to proceed up the Mohawk and assist General John Stanwix in the construction of what would become Fort Stanwix at the "great carrying place," the portage between the Mohawk River and Lake Oneida. This effort was designed to counter the threat of a French invasion down the Mohawk—the very operation that New France's governor general, Vaudreuil, had talked of earlier in the spring. It would also reestablish in the area a British presence that had been lacking since the fall of Oswego and would serve to renew Iroquois confidence in Great Britain. These objectives were openly discussed and waved about rather freely. Only Bradstreet seems to have known that he also had Abercromby's permission to proceed beyond these defensive actions and at long last take the battle to the gates of Fort Frontenac.[9]

Fort Frontenac, like most North American fortifications of this era, was not an impregnable citadel—far from it. But its strategic location where Lake Ontario empties into the Saint Lawrence River could not be denied. It was the crossroads of the western half of New France. Eastward, the Saint Lawrence funneled to it a singular lifeline of supplies and what sustenance Quebec and Montreal could spare.

Westward—ah, westward; that was the key. From Fort Frontenac westward, French supply lines and the accompanying influence of the French fanned out like water from a sprinkler. Indeed, Fort Frontenac was the faucet that dispersed goods to frontier posts from Fort Niagara and points south, including the Allegheny portage to Fort Duquesne, all the way westward to Fort Michili-mackinac. In return, the riches of New France's far-flung fur trade passed through Fort Frontenac and were funneled in the opposite direction down the Saint Lawrence to Montreal.

Count Frontenac, as the newly arrived governor general of New France, had established the fort in 1673 and named it for himself. Count Frontenac was eager to extend French influence on Lake Ontario by controlling its outlet, as well as to establish a line of some resistance that would both keep the Iroquois south of the lake and protect the French fur trade. Six years later, the French built Fort Niagara on the lake's inlet from the Niagara River and Lake Erie in an effort to extend this buffer. These efforts met with limited success, of course, and in time the British constructed the post at Oswego in response.[10]

Lulled by eighty-five years without a direct major attack, Fort Frontenac had grown far more commercial than military in nature. Arranged in the form of a square with bastions at the corners and stone walls about one hundred yards in length, its interior accommodated officers' quarters, barracks, a chapel, and the requisite powder magazine. Some thirty cannon graced the ramparts with another thirty held in reserve, but the fort's real treasure lay outside its walls in the great storehouses filled with goods at the water's edge. Frontenac was less militarily prepared than either Fort Niagara or Fort Duquesne, but each of these outposts owed its survival to the supplies that passed through Frontenac's wharves.[11]

As Bradstreet assembled his troops and headed up the Mohawk in August 1758, the makeup of his force was no secret. "Colonel Bradstreet is to command in an expedition this way of 3,000 men," noted a report from the Great Carrying Place dated August 13. His army was to include only 155 regulars and rely on provincials numbering as follows: "New York, 1,112; New Jersey, 412; Boston, Col. Williams, 432; Boston, Col. Doty's 243; Rhode

Island, 318"; and, of course, 300 or so of his fellow bateau men. Where they were headed, however, was still supposed to be a closely guarded secret. A train of eight artillery pieces and three mortars "go into Wood Creek this day" from the portage with the remainder of the army to follow the next, but their destination "is not known to any mortal here, except General Stanwix."[12]

That was probably not entirely true, particularly as a report a few days later made mention of very low garrison strength at Fort Frontenac and suggested that "they were not in the least suspicious of any army coming that way this year." It was possible, of course, that Bradstreet might strike westward and attack Fort Niagara instead, but regardless of the final destination, as his forces assembled and moved out, two things were certain.[13]

First, the Onondaga and Oneida from the Six Nations whom Bradstreet had encouraged to join in the expedition as allies quickly

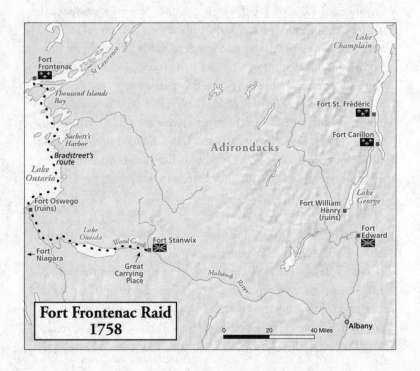

Fort Frontenac Raid 1758

became convinced that its purposes were far more aggressive than merely building a fort at the Great Carrying Place. Unwilling to carry the fight against posts where they frequently traded, most Indians quietly deserted the campaign. The second certainty was that the French, transfixed by operations at Louisbourg and Fort Carillon, were almost totally unprepared against any offensive operations on Lake Ontario. Not only had they left the back door largely unguarded; they had left it standing wide open.

So Bradstreet's bateau navy sped his little army down Wood Creek to Lake Oneida and then down Oswego Creek to Lake Ontario. Arriving there on August 21, they found only the ruins of the British post that Montcalm had captured and destroyed two summers before. The next morning, its destination now becoming obvious to all, the flotilla rowed thirty-five miles northward along the eastern shores of Lake Ontario to Sackets Harbor, where it paused to regroup. (Montcalm had used this quiet bay to similar advantage when he attacked in the opposite direction.)

Concerns ran high that any number of French vessels known to be on the lake might chance on the convoy of more than 100 bateaux and whaleboats and scatter it with cannon fire, but no French sails appeared. On August 25, after another twenty-five miles, Bradstreet's bateaux nosed ashore about one mile west of Fort Frontenac and landed their cargoes of troops, artillery, and supplies without opposition.

The French commandant at Fort Frontenac was Major Pierre-Jacques Payen de Noyan. He had spent forty-six of his sixty-three years in the service of his king in the French marines and was about to undertake his last official act. Forewarned several days earlier by his own Indian allies of Bradstreet's advance, Noyan had sent a messenger hurrying to Montreal to beg Governor Vaudreuil for any and all assistance. The governor immediately summoned militia from the fields where they were harvesting the crop that was New France's lifeblood and started them up the river toward Fort Frontenac. Meanwhile, Noyan mustered his forces and looked west from the fort's ramparts.

By the morning of August 26, Bradstreet's artillery was in place about 500 yards west of the fort. These pieces began a bombardment limited by the quantity of ammunition available. The French replied for show, but any real effect was limited on their side as well by a lack of troops to man all the fort's cannons. Unsure of how many troops Noyan had inside Frontenac, Bradstreet no doubt also had visions still fresh in his mind of Abercromby's frontal assault at Ticonderoga. Bradstreet did not order a direct attack. Instead, his troops advanced to an old line of defensive breastworks about 250 yards from the fort and then hauled two twelve-pounders to the top of a small hill only 150 yards from Frontenac's northwest bastion.

At dawn on the second morning, Bradstreet's artillery began to pour a pounding barrage against the fort's crumbling, eighty-five-year-old stone walls. "Being so near," wrote an officer in the New York regiment, "every shell did execution." The French cannon responded, but Noyan knew that he was doomed. There was no sign of relief from Montreal. Concerned for the large number of women and children in the fort and with no chance of overcoming Bradstreet's overwhelming odds in manpower, Noyan struck his colors. He lowered the white fleur-de-lis and ran up a red flag of surrender, which the French used to differentiate such a request from their national ensign.

Bradstreet wasted no time. Noyan's meager garrison—which totaled only 110 men at the time of the surrender—could keep their money and their clothes, but they were to return to Albany as prisoners of war until exchanged. The real prize, Bradstreet knew, was in the storehouses along the docks and in the holds of nine captured ships—two snows, three schooners, three sloops, and a brigantine.*

---

* This flotilla gives insight into the extent of French naval presence on Lake Ontario. Generally, a snow was two-masted and square-rigged; a schooner two-masted, rigged with fore and aft sails; a sloop single-masted, rigged with fore and aft sails and a jib; and a brigantine two-masted and squared-rigged except for fore-and-aft mainsails. Why these ships were allowed to cluster here when Noyan had warning of Bradstreet's advance is another matter.

Inexplicably, they had remained tied to the dockside rather than fleeing before Bradstreet's artillery could be hauled into position. Seeing the fort capitulate, the crews of the brigantine and a schooner attempted too late to get under way, but British cannon fire interrupted their escape. The French sailors quickly abandoned their ships and fled to shore in small boats. The brigantine and the schooner were left to drift, coming aground several miles away, and Bradstreet's men soon retrieved them.[14]

Not knowing when Governor Vaudreuil's Montreal militia might appear, Bradstreet's men quickly fell to work salvaging appropriate plunder before demolishing the fort. "It had in it a vast quantity of provisions, which we burned in the fort, and everything else that the French left, and we could not bring away," an officer from New York reported.

By one account, "2000 barrels of provisions"—flour, beef, and pork—went up in smoke. The two snows, the three sloops, and two of the schooners were also burned to their waterlines. Now, even if other provisions could be found, there was no way for the French to transport them to the western posts of Fort Niagara and Fort Duquesne. Even Bradstreet did not realize at the time how many of the hopes of New France were disappearing into the late summer sky amid the black smoke.[15]

By the afternoon of the following day, August 28, Bradstreet was anxious to be on his way. He amended the terms of surrender slightly and now permitted Noyan and his soldiers and their families to proceed directly to Montreal, on Noyan's promise to arrange for the release of a like number of British prisoners. This was not a noble gesture but merely Bradstreet's way of ridding himself of unnecessary baggage so that he might do what he did best: move fast. With all the plunder his men could gather loaded into the captured brigantine and schooner, Bradstreet ordered his fleet of bateaux and whaleboats back onto the lake.

"I have the pleasure to inform you," wrote one of his lieutenants just two days later, "that I arrived here [Oswego] this day, with a brig under my command, deeply loaded with furs, skins, bale goods, liquors, etc. The brig is the same that was taken from us by the French at Oswego two years ago."[16]

By the time that Governor Vaudreuil's militia finally reached the charred ruins of Fort Frontenac, Bradstreet had also arrived at Oswego and was once more hauling his wooden fleet back to the Great Carrying Place. His troops had suffered the loss of only one man killed and a dozen or so wounded.

Bradstreet immediately hurried on to Albany and pleaded with General Abercromby to allow him to conduct a similar raid against Fort Niagara. Had the bateau man known that the French garrison there numbered but forty men, he might have relied on the odds of a speedy victory, raided without Abercromby's permission, and then simply begged forgiveness. But now, having asked the question, Bradstreet was bound by the old plodder's answer. Abercromby—clearly not recognizing the strategic coup that Bradstreet had just wrought—said no. Niagara would have to wait for another year.[17]

Earlier that spring, Brigadier James Wolfe, who was so critical of so many, had already written of Bradstreet, apparently without ever meeting him, "Bradstreet for the bateaux and for expeditions is an extraordinary man." Now, he rendered his postmortem. "Bradstreet's coup," wrote the general who still dreamed of Quebec, "was masterly."[18]

Eighty miles north of Albany, Montcalm stood on a parapet at Fort Carillon as a chilly wind of early fall blew across Lake Champlain. It was now his turn to ponder his fate. The couriers of September 6 had not brought good news. In a double blow, Montcalm learned simultaneously of the defeats at Louisbourg and Fort Frontenac.[19] What good was his own victory now? Had it indeed been providence, Abercromby's stupidity, or just plain luck?

Replacing Lord Loudoun with General Abercromby may not have been one of Pitt's shrewdest moves, but it seems to have had one definite impact in the field. Loudoun had planned to move against Fort Carillon in late May—a time when both the lack of provisions and the melting snows would still preclude speedy deployment of French troops from their winter quarters in Montreal and Quebec. Montcalm did not march south with the bulk of Carillon's reinforcements until after the arrival of the first supply ships in late June. By the time Abercromby finally struck north, it was early July. Had the British appeared at Carillon's gates a month earlier, they would have found it in a state of readiness similar to that which had just greeted Bradstreet at Fort Frontenac.[20]

But no matter to whom he owed his victory, Montcalm had it. On the very evening that he received the news of Louisbourg and Frontenac, he set out for Montreal to confront the man he had come to feel was as grave a threat to New France as any invader—Governor Vaudreuil. Abercromby had blundered badly below Carillon's entrenchments, and this had cost him 2,000 casualties and another summer. But Vaudreuil had been caught napping on Lake Ontario, and that had cost him not only the vital provisions of Frontenac's storehouses but—more important—the lines of supply to both Fort Niagara and Fort Duquesne. Montcalm was determined to do all that he could to repair the damage.

At first glance, Bradstreet's raid seems to pale beside the grand siege of Louisbourg and the carnage at Ticonderoga. But closer scrutiny suggests that quite possibly this little-known affair might have been the death knell of New France. After all, it was not the loss of Louisbourg alone that threatened New France. With a more powerful navy, France might have saved Louisbourg or ended up trading it back and forth again as in the past. On the Lake Champlain frontier, New France's boundaries had remained essentially unchanged for four years running—admittedly, thanks in part to Abercromby's blunders. But Frontenac was different. It sat astride the main artery that sustained not only Niagara and Duquesne but also all of the upper Mississippi and Ohio valleys. It sat astride New France. Time would soon tell, but the bateau man had indeed taken a huge whack out of the trunk of the tree.

10

BRADDOCK'S
ROADS AGAIN

The four roads that Major General Edward Braddock was charged with hurrying British troops along in 1755 had led to varied results. As the summer of 1758 waned, Louisbourg, of course, was in Amherst's pocket; but the road north from Albany still stopped well short of Ticonderoga, which was held by the French. The thrust west to Fort Niagara had yet to materialize. That left the road Braddock himself had taken—west across the Appalachians to the forks of the Ohio. In the three years since Braddock's defeat on the Monongahela, there had not been so much as a tremor in the supremacy of the fleur-de-lis flying above Fort Duquesne. Along with his other grand plans for 1758, William Pitt decreed that this must change.

No one cheered more loudly at this prospect than colonials in Pennsylvania and Virginia. Although they would come to disagree bitterly over the route to be taken and the division of spoils afterward, Pennsylvanians and Virginians were united in their disdain toward this bastion of New France. In the aftermath of Braddock's defeat, raiding parties fanning out from Fort Duquesne had made their western frontiers run red with blood. At the very least, a more resolute successor to Braddock might have rallied the British retreat from the Monongahela at Fort Cumberland and provided some buffer. Instead, Colonel Thomas Dunbar had not stopped until he led his troops across half of Pennsylvania and declared his intention to go into winter quarters in Philadelphia even though it was only the end of July.

The result of Dunbar's flight was that the French and their Indian allies killed British settlers and burned British villages and outposts from the upper Susquehanna to the crest of the Blue Ridge, effectively pushing Great Britain's frontier eastward some 150 miles. In Pennsylvania, the towns of Carlisle, York, and Lancaster became armed camps filled with fleeing refugees. Much of the Juniata's valley and the Susquehanna's other western tributaries were emptied of inhabitants.

The upper Potomac fared no better. Passing through Winchester, Virginia, en route to Mount Vernon that summer, George Washington found the town jammed with settlers streaming eastward in a high state of anxiety. "Dunbar's decision to march to Philadelphia," Governor Horatio Sharpe of Maryland grumbled to Dinwiddie of Virginia, "has alarmed the frontier more than Braddock's defeat."[1]

Rumors of a French advance to the very outskirts of Philadelphia were merely that, but some uncertainty, fueled by countless skirmishes, continued unabated for the next three years. Throughout the early months of 1758, newspapers were still filled with widespread reports of brutal attacks, including scalpings. (These were by no means one-sided, as British and colonial troops were quick to visit similar atrocities on their attackers.) Pitt's three-prong strategy, bolstered by his reforms in recruiting and compensating provincial forces, at last provided a plan to restore some semblance of peace to this frontier. The British would march along the main fork of Braddock's roads again and strike directly against what most viewed as the chief source of all the trouble—Fort Duquesne itself.

To lead this third prong of his grand offensive of 1758, William Pitt chose a reserved but resolute Scotsman, John Forbes. He was born in Fifeshire in 1710 and initially trained as a doctor, but he discovered his real passion when he received a commission in the Scots Greys in 1735. He saw service in Flanders during the War of the Austrian Succession and was at Culloden along with—as it must seem by now—nearly every other officer of promise in the British army. By 1750, Forbes was a lieutenant colonel and, like both Amherst and Wolfe, had acquired considerable experience in logistics, serving for a time as the army's deputy quartermaster general. In the spring of 1757, Forbes was posted to America to serve on Lord Loudoun's staff as his adjutant general. When word of Loudoun's recall reached New York the following year, Forbes's good standing with Pitt and Lord Ligonier earned him advancement to brigadier general and command of the campaign against Fort Duquesne.

Initially, the bateau man Colonel John Bradstreet was assigned to Forbes as supply chief, but Bradstreet pleaded with Abercromby that he was an "utter stranger" to the Pennsylvania country—where, we might note, the rivers tended to flow across the general route west, not along it. Abercromby acquiesced and kept Bradstreet on the Mohawk. In his place, Forbes's quartermaster responsibilities were given to Colonel Sir John St. Clair, who had performed exactly the same services under General Braddock three years before. Whether St. Clair had in the intervening time gained any experience that would prove valuable remained to be seen.[2]

As his second in command, Forbes was to have Lieutenant Colonel Henry Bouquet. Bouquet was born in Switzerland of French Huguenots who had fled there from France to avoid religious persecution; he, too, had fought in the War of the Austrian Succession and risen to the rank of lieutenant colonel in the Prince of Orange's Swiss Guards. He was recommended for the Royal American Regiment and came to Pennsylvania in 1756 to recruit for this storied unit of British regulars raised in America from among largely German-speaking immigrants. Before his assignment with Forbes, Bouquet commanded a battalion of the Royal Americans on the North Carolina border.

Forbes and Bouquet made a good team. Each was every inch a

soldier. Already plagued by the poor health that would persist throughout the campaign, Forbes quickly became the expedition planner; and Bouquet became the executor of that plan in the field. Well aware of the hazards that had befallen Braddock's strung-out command, Bouquet quickly took wilderness warfare to heart and wrote his own recommended tactics for regulars marching through wooded terrain.

These were not exactly the rules of Rogers' Rangers, but they were a decided improvement over the standard drill manual. Secure lines of communication among all units, constant scouting of the line of march, and the occupation of "all suspected places . . . where ambuscades may be concealed" were absolutely essential, Bouquet theorized. "In case of an attack," he continued, "the men must fall on their knees; that motion will prevent their running away, and in covering them from the fire, shall give time to reconnoiter and to make the necessary dispositions."[3]

As he made his own assessment of the troops under his command, Forbes was determined, of course, that there be no "running away." He was glad to have Colonel Archibald Montgomery's regiment of Highlanders newly arrived from England and Bouquet's battalion of Royal Americans. When a company of artillery was included, these accounted for about 1,400 regulars. The remainder of his troops—5,000 or so—were to be provincial regiments, including the First Virginia under the command of George Washington.

A few provincial units had been in existence long enough to have acquired some measure of military discipline, but even Washington found plenty of frustration. Given the accustomed flow of liquor and rum, the young colonel declared in despair that it was difficult to maintain military discipline because of "the villainous behavior of those tippling housekeepers." Forbes was more blunt. Observing the new troops answering Pitt's most recent summons, the general complained bitterly to Pitt that "a few of their principal officers excepted, all the rest are an extremely bad collection of broken innkeepers, horse jockeys, and Indian traders, and that the men under them, are a direct copy of their officers" and a gathering from "the scum of the worst people."[4]

By which route would they march? Braddock's road from Fort Cumberland was choked with new growth and scented with the smell of defeat, but it still led to Fort Duquesne. Forbes seems to have initially assumed that this would be his route. It was certainly the obvious one. But as the quartermaster general, Sir John St. Clair, scurried about the Susquehanna Valley arranging teamsters and supplies for Forbes's advance, he came up with a different idea. Whether it was a result of persuasive lobbying by Pennsylvania merchants, a true eye for tactical military advantages, or simply a desire to avoid revisiting the grim path of 1755 is debatable. The result, however, was that St. Clair suggested discarding Braddock's route and carving a new road straight west through central Pennsylvania to Fort Duquesne.

By early May 1758, Forbes agreed and made plans for a methodical advance not from Fort Cumberland, but from tiny Raystown on the upper Juniata River. Although this decision regarding "which road" appears to have been made easily, it would have important ramifications far beyond the present military tactics.[5]

For one thing, young Colonel Washington was apoplectic. When Washington received an order to concentrate his First Virginia regiment at Fort Cumberland and then cut a thirty-mile road north to Raystown instead of preparing the way back over Braddock's road, he could not believe it. How much of Washington's dismay sprang from military concerns and how much was the result of whispers in his ear from Virginia's own commercial interest, has long been debated—particularly by those who cannot imagine Washington acting with anything but the most noble of purposes.

Virginia, however, had long claimed lands in Ohio. Washington had been in the forefront of those claims. Both Virginia and Pennsylvania were only too aware that once the present conflict was resolved, a new wave of settlers would flow westward via whichever route had been taken. This would not only boost the economy of the colony of its origin, but also benefit it defensively and commercially with the new string of outposts to be built along it. Now, Forbes was determined to do just that through the center of Pennsylvania. On June 24, 1758, Colonel Bouquet's advance troops reached Raystown, where they paused for the better part of a month to build Fort Bedford.[6]

By the end of July, Washington met Bouquet in Raystown and protested against the choice of routes almost to the point of insubordination. Bouquet, a proper Swiss officer, was not used to having orders questioned, particularly by a twenty-six-year-old provincial officer whose military record to date had been largely one of defeat. Washington found the colonel determined to pursue "a new way to the Ohio; through a road, every inch of it to cut, at this advanced season, when we have scarce time left to tread the beaten tract; universally confessed to be the best passage through the mountains." Meanwhile, Colonel John Armstrong of Pennsylvania observed: "The Virginians are much chagrined at the opening of the road through this government [Pennsylvania], and Colonel Washington has been a good deal sanguine, and obstinate upon the occasion."[7]

By early August, Bouquet had 600 Pennsylvanians, who were guarded by another 600 Virginians, hacking the way westward from Raystown up the main spine of the Alleghenies. Sir John St. Clair, who had pushed so hard for this route, had already had second thoughts and decided that Fort Duquesne could not be reached that year without "going into Braddock's old road." Hardly beyond the smoke of Fort Bedford, St. Clair reversed himself again and urged Forbes to "send me as many men as you can with digging tools, this is a most diabolical work, and whiskey must be had."[8]

Whatever his misgivings and constant supply shortages, St. Clair soon reported to Bouquet that he had examined the early miles of the road west from Fort Bedford and pronounced them "good." Bouquet passed this news on to the still disgruntled

Washington and could not refrain from twisting the barb just a little. " I cannot therefore entertain the least doubt," Bouquet wrote to Washington, "that we shall now all go on hand in hand and that the same zeal for the service that has hitherto been so distinguishing a part of your character will carry you by Raystown over the Allegheny Mountains and on to Fort Duquesne."[9]

But for the moment, while Washington continued work on the road from Cumberland to Raystown, it was Bouquet who led off again, pushing the line across the main spine of the Alleghenies and Laurel Hill a total of forty miles farther west to Loyalhanna Creek. Here, his troops paused to build Fort Ligonier and await both Forbes with the remainder of his command and St. Clair's snaking supply trains.

Suffering from what was probably dysentery, Forbes did not arrive in Raystown until September 15, in a litter strapped between two horses. Meanwhile, St. Clair had managed to antagonize half of Pennsylvania, but supplies were finally moving. His efforts were aided when the Pennsylvania assembly as late as September 20 authorized a bonus for anyone who would furnish four good horses and a wagon and haul at least 1,400 pounds of supplies, plus subsistence for the team, from Lancaster to Raystown.[10]

Fort Ligonier was within forty miles of Fort Duquesne, and Bouquet was caught between pressing onward and waiting for Forbes. Part of his quandary was not knowing the strength of Fort Duquesne's defenses or the number of its garrison. Actually, the "impregnable fortress" of Duquesne existed far more in the British imagination that it did in fact. Even in the immediate aftermath of Braddock's defeat, its commandant had admitted that "if the enemy had returned to the attack with the thousand fresh troops they had in reserve the defenders of the fort would perhaps have been 'seriously embarrassed.'"[11]

Three years later, that still held true. Of the sorry state of the fort's defenses, Jean-Daniel Dumas, who had held the field against Braddock's advance, wrote that Fort Duquesne was "fit only to dishonor the officer who would be entrusted with its defense."

Now, in August 1758, that officer was François-Marie le Marchand de Lignery and even he couldn't be certain of the exact number of troops under his command because of the free flow of Canadian militia and Indian allies into and out of his gates. Estimates ranged from 1,000 to more than 2,000, but some of these were no doubt strung along the Allegheny supply line from Presque Isle.[12]

With Bouquet strengthening his forward base at Fort Ligonier, Lignery no longer had any doubt as to which route the British were taking. Whatever his numbers, he lacked the manpower to force a confrontation in the open. And, with dwindling supplies made even sparser by Bradstreet's raid on Frontenac, there was no way that he could survive a protracted siege. His only alternative was to make a series of rapid raids designed to stall the British at Fort Ligonier and hope that the fall rains would mire them in their tracks.

In the initial Indian raids against Fort Ligonier in early September, the attackers captured several British soldiers, killed and scalped a Highlander, and caused enough anxiety among the British rank and file that Bouquet weighed the option of sending out small raiding parties in response. Then, Major James Grant of the Highlanders proposed a bolder strategy. Give him 600 men, Grant offered, and he would lead a reconnaissance in force, not only roughing up the Indians a bit, but also harassing the fort and its supply lines, perhaps even putting it under siege. Bouquet chose to believe an erroneous report that gave Fort Duquesne's garrison strength as only 600 men, and he sent Grant westward with 400 regulars and 350 provincials.

By September 13, 1758, Grant's detachment had moved to within five miles of Fort Duquesne, and Major Andrew Lewis led about 400 men in an attack against some of the fort's outbuildings. The French certainly knew that the British were in the vicinity in force but made no attempt to sally forth to oppose them. Before dawn the next morning, Grant divided his meager command into three columns and prepared to move against the main French encampment. (More than a century later, a certain Colonel Custer would order a similar division of command with even more disastrous results.)

Lewis, 100 regulars, and 150 Virginia troops went first but ended up stumbling around in the dark and had to return to regroup. Somewhat disgusted with this first effort, Grant sent Lewis and these troops to lie in ambush near his baggage train should the French attempt to attack it. Captain William MacDonald and 100 Highlanders were then given the unenviable task of marching toward the gates of Fort Duquesne with drums beating to act as bait to lure the French out of the fort and into the open field. Grant and his remaining 400 troops waited in ambush on the nearby hills.

The French did come out of the fort, but not with the result Grant had intended. Perhaps as many as 1,000 French and Indian

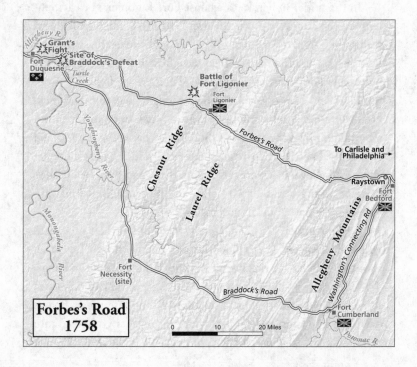

Forbes's Road
1758

warriors swarmed to the attack. "The 100 Pennsylvanians who were posted upon the right at the greatest distance from the enemy, went off without orders and without firing a shot." The French spread out and soon were counterattacking furiously.

"We were fired upon from every quarter," Grant admitted. He attempted to fall back to the baggage train, where he supposed Lewis was. Lewis heard the firing and tried to come to Grant's aid, but his advancing troops took a different path and now both units were subjected to French cross fire. MacDonald's Highlanders were cut off completely. By the time the engagement was over, Major Lewis and Major Grant, along with a large number of their troops, had been forced to surrender. Only Captain Thomas Bullit's unit of about 100 Virginians held the line and "sustained the battle with all their forces." Without them, there might have been a complete rout.

As it was, Grant's battered force limped back to Fort Ligonier without him after losing almost 300 killed or captured. Once again, a Highlander regiment bore the brunt of the British losses. French losses were eight killed and eight wounded. In many respects, it was as one-sided a victory as the French had achieved over General Braddock three years before. The main difference was that instead of turning tail and running halfway across Pennsylvania, the British were able to fall back only forty miles to Fort Ligonier and regroup. Forbes's strategy of advancing in force with heavily armed bases of support was paying off.[13]

The immediate result of Grant's blunder was that it emboldened the French to launch a counteroffensive against Fort Ligonier. Fall was well along, and another bloody nose here might still send the British back over the mountains and buy Fort Duquesne another winter.

On the morning of October 12, 400 French regulars and militia along with about 100 Indian allies skirmished with British units outside Fort Ligonier. Forbes was still en route from Fort Bedford, and Bouquet was off inspecting portions of the road, but Colonel James Burd commanded the counterattack. After two hours of heavy fighting the French were forced to withdraw, but their Indian allies made off with the majority of the British herd of horses. The entire encounter proved brief, with minor casualties on either side; but when Montcalm heard of the affair, he thought that the French had gained "a considerable advantage." In reality, it should have been a warning for them. Forbes's plodding advance would not be brushed off lightly or sent back down its road as Braddock's had been.[14]

Forbes's military strategy of a heavily fortified advance was working, but now there occurred an event of strategic importance every bit as critical to Forbes's advance as the hewing out of a road. As long ago as the Albany Congress of 1754, Sir William Johnson had been the designated point man for British negotiations with all Indian nations. While this had certain advantages, it presupposed a commonality among Indian nations that simply did not exist. This was particularly true of relations between the Iroquois, by whom Johnson was held in high esteem, and the Delaware, by whom he was not. This was in large measure because the Delaware strongly opposed efforts by the Iroquois to dictate Delaware policies toward France and England just as they did for the members of the Six Nations. Knowing that support from the Indians would be crucial to his advance, and remembering that Braddock's lack of it had contributed his downfall, Forbes received authority to bypass William Johnson and negotiate directly with the western Delaware, Shawnee, and other nations sometimes collectively referred to as the "Ohio Indians."

Because Johnson had a history of holding "his" Iroquois close to the New York campaigns, Forbes first sought to persuade the Cherokee and Catawba allies in the south to join his expedition. Aided by his cousin James Glen, a former governor of South Carolina who had long fostered close ties with the Cherokee, Forbes was able to recruit upwards of 700 warriors. These arrived in Pennsylvania in mid-May 1758 and were loosely attached to William Byrd's Second Virginia Regiment. The problem was that at this time of year and given Forbes's strategy of a plodding advance, there was not much for them to do. Despite Forbes's protestations that he had gone to great lengths with presents to appease them, he found that "nothing will keep them." The Cherokee and Catawba soon became disaffected by the incessant delays and dispersed to their southern homelands.

Having failed to muster large numbers of his own Indian allies, Forbes continued his diplomatic efforts to deny the French the use of theirs. First, he urged the Pennsylvania legislature to honor its commitment of 1757 to Teedyuscung and his eastern, or Susquehanna, Delaware to reserve lands in the Wyoming Valley for them. Forbes then used a willing Teedyuscung as an emissary to the western Delaware and Shawnee. Teedyuscung was instrumental in arranging for a Moravian missionary, Christian Frederick Post, to visit the Ohio Indians in late August 1758. "Why do not you and the French fight in the old country, and on the sea?" they asked Post. "Why do you come to fight on our land? This makes everybody believe, you want to take the land from us by force, and settle it."[15]

Post had no good answer to that, of course, but he could tell them of Pennsylvania's promise to Teedyuscung and of Forbes's dogged advance. The latter became even more obvious when Colonel Bouquet did not abandon Fort Ligonier despite Grant's defeat and the French attack on Ligonier itself. The result of Post's journey was that delegates from Teedyuscung's eastern Delaware and the Ohio Indians met with representatives of Pennsylvania in Easton, north of Philadelphia, in October 1758. Also in strong attendance were members of the Six Nations who were determined to reassert their feudal control over the external policies of these nations. Teedyuscung may have been the mediator, but faced with the appearance of the Iroquois, he quickly lost any role as a power broker.

The outcome of the conference in Easton was twofold. First, the Iroquois demanded that the Pennsylvanians renounce their purchases, resulting from the Albany Congress, of all lands west of the Alleghenies. These included country around the forks of the

Ohio long claimed by the Iroquois through their influence over the Ohio Indians. Pennsylvania's agents agreed, and their acquiescence effectively recognized Iroquois hegemony over the other nations in the region, particularly as the lands were symbolically returned directly to the Iroquois, not to the Ohio Indians.

Second, however, the Delaware and Ohio Indians were appeased by a promise from Governor William Denny of Pennsylvania that despite whatever influence the Iroquois might have over them generally, Pennsylvania would continue to deal directly with them on matters of local concern. These were essentially a respect for past treaty reservations and a resumption of trade once the French were expelled.

This left all parties with most of what they wanted. The Iroquois had reasserted their dominance over the Ohio Indians in external affairs; the Ohio Indians and eastern Delawares had received certain assurances of territorial integrity from the Pennsylvanians; and the Pennsylvanians had won the Ohio Indians and eastern Delawares back into the fold of British influence.

Arguably the most important Indian conclave in Pennsylvania's history, the Treaty of Easton was formalized on October 25, 1758. When word of it spread westward and reached Fort Duquesne, those Ohio Indians who had been allies of the French for more than three years quickly melted away into the forest. Suffering from a lack of supplies and now deserted by their Indian allies, the remaining French soldiers at Fort Duquesne were feeling increasingly isolated and alone as General Forbes, still in agonizing pain from his illness, considered his next move.[16]

As Forbes gathered the bulk of his forces at Fort Ligonier, time was of the essence. The other campaigns of 1758 had long since come to a close. Louisbourg was settled in for a winter under a British garrison. Abercromby and Montcalm were gone from the waters of Lake Champlain and Lake George. Bradstreet was back in Albany after his romp to Frontenac. With the skies of November darkening, only Forbes and his command were still in the field. Now, with some 5,000 troops assembled and the Ohio Indians neutralized, it was time for one final effort against the longtime French thorn at the forks of the Ohio.

In the British assemblage was the First Virginia Regiment. Its colonel was still complaining about the state of the Pennsylvania road, calling it "indescribably bad" and voicing his view that the campaign would grind to an end at Fort Ligonier.[17] Forbes thought differently, however, and dispatched Washington and about 500 Virginians west in pursuit of yet another French raiding party that had stolen horses from near the fort. Perhaps thinking of Grant's misfortune, Forbes quickly sent another Virginia unit under Colonel George Mercer on his heels. Washington was successful in capturing three prisoners: an Indian couple and an Englishman who claimed to have been recently kidnapped from Lancaster.

Just as Washington was interrogating them around a campfire, Mercer's troops appeared on the scene and somehow surmised that it was an enemy encampment. In the darkness, a furious exchange of friendly fire occurred that left two officers and thirty-five enlisted men dead. Only the intervention of Captain Thomas Bullit, hero of Grant's last stand, frantically waving his hat in the midst of the melee, prevented further carnage. Washington made no contemporary mention of the unfortunate incident, although it was reported in the *Pennsylvania Gazette* and by Forbes to his superior. The only good thing to come out of the encounter was that the captured Englishman proved of dubious loyalty and was in fact serving the French out of Fort Duquesne. It didn't take much frontier persuasion to make him give a full accounting of its current condition and declining garrison. Forbes was as pleased as his weakened physical condition would allow.[18]

For their part, Fort Duquesne's French defenders were glum. Even if they could hold out militarily against Forbes's advance and recruit approaching winter as their ally, they couldn't survive a siege. There simply was no food, or other supplies, to permit them the time, no matter how gallant their intent. And there was no food coming. What succor there might have been had gone up in flames in the storehouses of Fort Frontenac a few months before. Now, the extent of Bradstreet's raid really became clear. The trunk had been whacked, and the leaves were starting to fall from the tree of New France as surely as they were falling from the oaks and maples of the surrounding forests. "I am in the saddest situation one could imagine," commandant Lignery lamented to Governor Vaudreuil.[19]

So Forbes sent Washington's Virginian regiment and John Armstrong's Pennsylvanians west from Fort Ligonier to hack out the remainder of his road. Finally, Washington seems to have embraced the road building. The units crossed Chestnut Ridge and pushed on to Bushy Run and then the upper reaches of Turtle Creek. Only about a dozen miles from Fort Duquesne, they established Bouquet's Camp and waited for the arrival of General Forbes and the remainder of his little army.

According to an eyewitness account in the *Pennsylvania Gazette*, the following evening "a heavy firing was heard from thence [Fort Duquesne]." It "continued first for about an hour, then ceased for some time, and began again, and lasted half an hour; and that afterwards a rumbling noise was also heard, like that of great guns at a distance." No one was quite sure what to make of it at the time, but an investigation the next day revealed the source of the huge explosion. As an express messenger to the *New York Gazette* reported: "Monsieurs did not stay for the approach of our army, but blew up the fort, spiked their cannon, threw them into the river, and made the best of their way off; carrying with them everything that was valuable, except the spot of ground where the fort stood."[20]

Indeed, Commandant Lignery had followed his orders and blown up the principal works of Fort Duquesne rather than let them fall into British hands. He and his remaining garrison of about 400 men dispersed, a few going down the Ohio toward the Illinois country and the others withdrawing up the Allegheny to Venango and Presque Isle. "After much fatigue and labor," wrote another British participant, "we have at last brought the artillery to this place, and found the French had left us nothing to do, having on the 24th instant blown up their magazines, and burned their fort to the ground."

There were grisly scenes to be encountered. The unburied bodies of Grant's fallen troops littered the ground for three miles to within 100 yards of the fort. By one account, the heads of a number of brave Highlanders were placed on "a long row of stakes that the Indians had erected along their well-beaten path to the fort . . . and underneath these were hung their Scottish kilts."[21]

As Forbes and his troops assembled at the forks of the Ohio and hurriedly began work on a small stockade to accommodate a winter garrison, a number of Ohio Indians, who only a few weeks before had been on the other side, gathered to confer with the general. His subordinates assured them that it would be a small facility, intended to reestablish trade. Afterward, the *Pennsylvania Gazette* reported "that the French, by being obliged to abandon Fort Duquesne, have lost a vast tract of country, and the various tribes of Indians inhabiting it, seem, in a certain manner, reconciled to his Majesty's protection and government."[22]

That hadn't been the bargain at Easton, of course, and the explosion that rocked Fort Duquesne did far more than signal the withdrawal of the French from the upper Ohio. The Treaty of Easton was scarcely six weeks old, but it was quickly forgotten. Give up British claims to the Ohio country? That was almost unthinkable. If Pennsylvanians and Virginians would argue over a road, they would certainly fight over the spoils of its destination, and they would certainly not abandon them.

The *Pennsylvania Gazette* of December 28, 1758, was quite straightforward about the matter. The recent campaign was not merely a struggle between France and Great Britain. At stake was "a vast country, exceeding in extent and good land, all the European dominions of Great Britain, France, and Spain, almost destitute of inhabitants; and will, as fast as the Europeans settle, become more so of its former inhabitants." A letter from a correspondent on the scene at Fort Duquesne was more succinct but no less sure of the result: "Blessed be God, the long looked for day is arrived that has now fixed us on the banks of the Ohio." The letter was dated November 28, from "Pittsburgh."[23]

Leaving a small garrison of Pennsylvanians to guard the hard-won forks of the Ohio through the winter, General John Forbes retraced his road and arrived back in Philadelphia on January 17. Part diplomat to the Indians, part military tactician, he had managed to build his own road in order to complete Edward Braddock's. Forbes would be called the victor of Fort Duquesne, but at least some of the laurels were due to John Bradstreet, who, with his raid on Frontenac, had cut the fort's umbilical cord.

Perhaps Forbes's real victory was that he negotiated in good faith with Teedyuscung of the Susquehanna Delaware and with the Ohio Indians and did not live to see his efforts repudiated. Shortly after his arrival in Philadelphia, while in excruciating pain, Forbes penned several letters to General Jeffery Amherst, who by now was Abercromby's replacement as commander in chief in North America. Forbes implored Amherst not to take the Indian alliances lightly. If British influence was to continue at the forks of the Ohio, Forbes told the new commander, relations with the Ohio Indians would have to be "settled on some solid footing, as the preservation of the Indians, and that country, depends upon it." Forbes bluntly recognized that "the jealousy subsisting between the Virginians and Pennsylvanians" over trade and lands was at the core of the problem and urged Amherst to exert strong leadership with both sides.[24]

Six weeks later, General John Forbes was dead at age fifty-one. It could quite rightly be said that he had given his life to follow Braddock's road. Now, wherever the roads would lead in the future, they would all belong to Jeffery Amherst. No one could say that he had not been forewarned.

As governor of Massachusetts, William Shirley forged a model of cooperation between Great Britain and its colonies that soon evaporated.
Massachusetts Historical Society

George Washington in the uniform of a Virginia colonel, a line engraving from a 1770 painting by Charles Wilson Peale.
Library of Congress, LC-USZ62-99148

"Join, or Die," often considered America's first political cartoon, from the *Pennsylvania Gazette*, May 9, 1754.
Library of Congress, LC-USZ62-9701

An engraving showing Major General Edward Braddock being carried
from the carnage of the battle of the Monongahela, July 9, 1755.
Library of Congress, LC-USZ62-63135

John McNevin depicted a much-too-young-looking Edward Braddock
when he envisioned Braddock's burial scene a century after the events.
Library of Congress, LC-USZ62-50571

With a wave of his hat in the surf off Louisbourg, Brigadier General James Wolfe set in motion Great Britain's conquest of Canada.
Library of Congress, LC-USZ62-48404

This medieval-style image of Major General Jeffery Amherst, painted after his return to England, suggests a conqueror far more at peace than was actually the case.
Library of Congress, LC-USZ62-45182

This British officer's rendition of a besieged Louisbourg,
looking out from Lighthouse Point, shows the town surrounded
and the French fleet bottled up in the harbor.

Library of Congress, LC-USZ62-2771

Harry A. Ogden depicted what may have been the high-water
mark of New France: Montcalm congratulating his troops after
the battle of Fort Carillon, July 8, 1758.

Fort Ticonderoga Museum

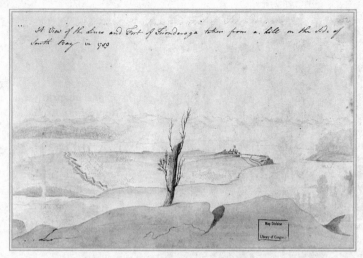

A View of the Lines and Fort of Ticonderoga taken from a hill on the side of South Bay in 1759

This simple drawing of Fort Carillon (Ticonderoga) and
the French lines from Rattlesnake Hill argues for the use
of artillery and not a frontal assault by infantry.
Library of Congress, LC-USZ62-133910

Robert Dowling painted the gallant but futile charge
of the Highlander Regiment against the French abatis
west of Fort Carillon on July 8, 1758.
Fort Ticonderoga Museum

Louis Joseph, marquis de Montcalm-Gozon de Saint-Véran, patiently defended Fort Carillon but then impulsively rushed into the field on the Plains of Abraham.

Library of Congress, LC-USZ61-239

Pierre François de Rigaud, marquis de Vaudreuil, governor general of New France, 1755–1760, was born in Canada but spent his last years under a cloud in France.

Library of Congress, LC-USZ62-110256

Sir William Johnson, circa 1756,
is shown looking more his role as
a proper British aristocrat than
as Indian commissioner.
Library of Congress, LC-USZ62-2695

Major Robert Rogers of the
Rangers: there was plenty of
adventure, but in the end
the facts did not live up
to the legend.
*Library of Congress,
LC-USZ62-45269*

King George III of Great Britain
and Ireland: "I glory in the name of
Briton," he intoned, but promptly
made a mess of things.
Library of Congress, LC-USZ62-96229

This view of the action at Quebec first appeared in *London Magazine* in 1760 and shows the landing at Anse au Foulon and the subsequent battle on the Plains of Abraham.

Library of Congress, LC-USZ62-47

It did not happen quite so poetically, of course, but a dozen years after the event, Benjamin West painted his version of Wolfe's final moments at the battle of Quebec.

Library of Congress, LC-USZ62-111

11

CARIBBEAN GAMBIT

Despite the expanse of geography that was at stake in Canada and the extraordinary lengths that Pitt's empire had recently undertaken to subject it, there was a region of North America that in terms of trade made Canada appear a pauper. This was the Caribbean. The days were long since past when the Caribbean could be called a Spanish lake, and the collision of empires resounded here just as it did throughout the remainder of the continent. With little regard to the indigenous peoples and even less to the hundreds of thousands of black slaves imported here from Africa, the European powers traded Caribbean islands and maneuvered for position as if pushing pawns on a global chessboard.

In some respects, the map of the Caribbean when the war was formally declared in 1756 was as complex and confusing as were the changing boundaries and alliances of Europe. Generally, Spain claimed Cuba, the eastern half of Santo Domingo, Puerto Rico, and Trinidad. Great Britain was established in Jamaica, the eastern Virgin Islands, Barbuda, Antigua, Saint Kitts, Nevis, Montserrat, and Barbados. Denmark held the Danish West Indies, or the western Virgin Islands of Saint Thomas, Saint Croix, and Saint John. The Dutch maintained a small but valuable trading presence from the island of Saint Eustatius, east of Saint Croix. This left France with the western half of Santo Domingo—now called Haiti and ceded by Spain to France by the Peace of Rijswijk of 1697—and Guadeloupe,

Martinique, and Grenada. The bulk of the mainland circling the Caribbean from Florida counterclockwise to Venezuela belonged to Spain, save for British interests in Belize and Dutch claims to Aruba and Curaçao. It was indeed a chessboard with multiple players.

Add to this patchwork the so-called neutral islands of Dominica, Saint Lucia, Saint Vincent, and Tobago. Except for Tobago, these were essentially what are now called the Windward Islands between Martinique and the coast of South America. Ostensibly, Great Britain and France agreed in 1730, and reaffirmed at Aix-la-Chapelle, that these "neutral" islands would not be colonized by either nation pending the settlement of rival claims to them. This may have made diplomatic sense, but the geography of the area produced a different result.

Dominica lay in the chain between French Guadeloupe and French Martinique; Saint Lucia and Saint Vincent between Martinique and French Grenada. Despite the proximity of the British at Barbados, about 100 miles east of Saint Vincent, it was only a matter of time before French settlers emigrated from nearby islands. This was particularly true of Dominica, where the French built plantations and subjugated the natives. By the time of the war, the French numbered 2,000 on the island and had 6,000 acres under cultivation by 6,000 black slaves.[1] So much for neutrality.

The long-standing feud between Great Britain and France over the neutral islands did not precipitate war, but it was certainly another log on the fire, particularly because the British came to view continued French settlement there as evidence that the French could not be trusted in any diplomatic matter. Far more than geography was at stake. The real value came from sugarcane. It was the cash crop of the Caribbean, cultivated by cheap slave labor and highly valued in both Europe and the remainder of the North American colonies.

France's wealth from sugar was centered on western Santo Domingo, which the French called Saint Domingue; Guadeloupe; and Martinique. Guadeloupe alone was said to produce more sugar than all of the British West Indies combined. Martinique produced less sugarcane, but its other crops, including coffee, averaged over 1 million pounds sterling annually. Even more important, Martinique was the residence of the governor general of the French Caribbean and the seat of its Superior Council, responsible for managing all French possessions in the West Indies, including Saint Domingue. Fort Royal on the island's western coast was the principal French naval base in the Caribbean. In short, Martinique was a ripe plum to be picked.

Conditions here were certainly much different from those along the Saint Lawrence River where Montcalm and his soldiers, as well as Canada's residents, were reduced to rationing flour by the ounce. In hindsight, one wonders how Louis XV and his ministers could have been so shortsighted in their support of New France; but it wasn't just the continental war in Europe that distracted them—it was the lucrative sugar trade from the Caribbean. Canada's exports of furs paled against the sugar exports of the Caribbean.

By one account, the trading needs of the French West Indies required some 1,400 merchant vessels. "The value of these islands when compared with Canada as a source of wealth to France," according to the historian Lawrence Henry Gipson, "may be measured by the fact that during the year 1754 but forty-one ships ascended the Saint Lawrence, including nine from the Caribbean Sea." It was clear that France's Caribbean trade was a source of income to be zealously guarded while Canada was decidedly on the short end.[2]

Before the war a key component of French trade in the Caribbean had been trade with the British colonies in North America. They were as eager as Europe for sugar, but they were also ready outlets for molasses, a by-product of the sugar mills that was essential to distilling rum. Given their preference for wine and brandy, the French actually forbade rum imports into France from the Caribbean, but the English colonies had no such compulsion and gladly traded foodstuffs and tobacco in return.

The outbreak of hostilities disrupted this mutually beneficial trade and gave colonial shipowners from Boston to Charleston pause. What were they to do now? Many turned a blind eye to British laws and continued their trading routines as smugglers. Others simply armed their vessels and became privateers, preying on French merchantmen. The French did likewise, of course, and British merchantmen also fell victim to French privateers.

As the numbers of ships lost on both sides escalated after the official declaration of war in 1756, both Great Britain and France began to use a convoy system similar to what the Spanish had used since the days of their first treasure fleets. French merchantmen assembled with warships at Cap François or Port-au-Prince on Saint Domingue. The French squadron from Fort Royal escorted merchantmen out of Guadeloupe, Martinique, or the "neutral" island of Dominica.

Similarly, the British rendezvoused at their naval bases at Port Royal in Jamaica and English Harbor in Antigua. In 1757, Great Britain's Jamaica convoy numbered 120 ships; in 1758 it numbered 164. But such military protection could not keep the vagaries of the weather from disrupting the passage, as a storm did in 1758 when ninety Virginian tobacco ships were scattered and left at the mercy of French privateers. According to Lloyd's, "between June 1756 and June 1760, a grand total of some 2,539 British and colonial vessels fell victim to French raiders on the high seas."

During this same period, the French lost 944 ships to the British, but this news was not as good as it might first appear. France simply had fewer ships to lose. By 1758, thanks to the war and despite its losses, British trade was booming. Mr. Pitt was indeed looking like an empire-builder. Shipyards throughout the British Isles and the North American colonies were sending ship after ship down the ways to replace the losses. Shipbuilding in France lagged far behind, and the French merchant marine simply could not keep up. "The French merchant marine," wrote the admitted Anglophile Lawrence Henry Gipson, "was being inexorably swept from the seas." Increasingly, France turned to neutral countries, particularly the Netherlands, in an effort to maintain its overseas commerce.[3]

France's use of neutral carriers to circumvent Great Britain's growing dominance of the seas led British admiralty courts to set down two famous rules of international law: the Rule of 1756 and the Doctrine of Continuous Voyage. Under long-established principles of international law, a truly neutral ship and its cargo could not be seized by one belligerent simply for trading with its enemy. The Rule of 1756 decreed, however, that if a country, such as France, treated a ship's cargo with respect to duties and taxes as its own, then the ship that carried it would be subject to seizure regardless of its flag.

In response, French ships began depositing Caribbean sugar and other products at the Dutch West Indies ports of Saint Eustatius and Curaçao for transport to France in neutral Dutch vessels with appropriate documentation. Great Britain responded: "but a voyage begun on a bottom that would render the cargo confiscable," that is, on an enemy ship, could not to be continued by a ship of a neutral as a continuation of the same voyage without still being subject to seizure. This was the Doctrine of Continuous Voyage. British colonists in North America cheered both these legal pronouncements which gave their privateers more possible maritime targets. Fifty years later, however, as the young United States, they would come to abhor the decisions when, during the Napoleonic wars, their attempts at neutral trade with France became one of the causes of the War of 1812.[4]

The Dutch, of course, were incensed, but there was not much that they could do militarily. Considering England's past ties with the house of Orange, it was slap in the face, especially in December 1758 when Pitt responded to the personal protests of the Dutch ambassador in no uncertain terms. According to a later report from The Hague, Pitt railed "that the [Dutch] navigation and trade to the French islands were carried on for the account of the French though under borrowed names" and that all the certificates professing Dutch ownership were "false and counterfeit." In addition, Pitt asserted, "the merchants concerned in that trade preferred gain to their eternal salvation; and that by false oaths they had given up their souls to be eternally damned."[5]

What France desperately needed at this stage of the war was a powerful maritime ally. Clearly, the Dutch could not fill the bill. The only country that could was Spain, related to the French throne by the Bourbon blood of Louis XIV, but still intractably neutral at the insistence of his great-grandson Ferdinand VI. At the time of the formal declaration of war between Great Britain and France in 1756, the French navy had sixty-three ships of the line. Spain had forty-six ships of the line. Taken together, a combined French-Spanish fleet would have numbered 109 capital ships. By comparison, in 1756 the Royal Navy numbered 130 ships of the line.[6]

Disregarding the state of readiness of each vessel—which varied greatly in fact and in the telling—and remembering that ships of the line were but one measure of naval strength, the combined French and Spanish navies would have at least enjoyed some measure of parity with the navy of Great Britain. It is one of history's "what ifs," but it was not going to happen—at least not in 1756 or, for that matter, two years later, in 1758. Spain was neutral.

"We shall leave our readers to judge what it indicates," reported the *South Carolina Gazette* from Charleston in July 1758, "But it is certain, that the Spaniards have now a very strong garrison at Saint Augustine; and are building with all possible dispatch, a strong fort at the southwest of the town."[7] What this indicated was that such reports made for nervous colonials. The reports did not, however, change diplomatic fact. Spain was neutral.

A subsequent letter from Saint Augustine reported that "no more than 40 dragoons" had arrived there with the new governor from Havana; that no new fortifications were being erected; and that the entire garrison of Spain's stronghold in Florida numbered about 450 men. More to the point, "all French privateers were forbid the Port of Saint Augustine; and that by advises received from

Spain as late as June, it appeared that his Catholic Majesty was not the least disposed to enter into, or be concerned in the present broils in Europe or America."[8] So Spanish neutrality seemed an undisputed fact, as Pitt's global strategy brought his gaze to bear on the Caribbean and in particular the sugarplums of Martinique and Guadeloupe.

One source, which appears to have been quoted many times without careful scrutiny of its veracity, suggests that as many as 1,400 of the British ships captured by French privateers were brought to the island of Martinique.[9] Militarily, that should have given Pitt ample incentive to launch an attack against Fort Royal, but it was actually the elixir of sugar that turned his head in this direction. More precisely, it was the powerful sugar lobby in Great Britain. Pitt's good friend, sugar magnate William Beckford, didn't come right out and argue the benefits that would accrue to his fellow sugar planters should the capture of Martinique result in more product or less disruption in trade. Rather, Beckford suggested that Martinique might be captured and then exchanged to France in return for the loss of Minorca, which Great Britain still considered essential to its power in the Mediterranean.

After the previous war, Louisbourg had served as a similar bargaining chip. Now, Beckford lobbied Pitt that because Louisbourg was the key to Canada, it must be retained at all costs and another chip acquired. It is debatable whether Beckford really believed that, or was merely repeating Pitt's party line, or was prescient enough to know that the American colonies would howl in even greater indignation should Louisbourg be given up again. Whatever his reasons, as long as Louisbourg could not be the bargaining chip, a move against Martinique was all the more valuable. On September 11, 1758, Beckford wrote to Pitt a second time, with greater urgency. Noting that the island "has but one town of strength," Beckford argued that the "conquest [was] easy" and that Pitt should "attempt it without delay and noise."[10]

Truth be told, Martinique was having a spell of bad luck. Not only was the war taking a toll on its merchantmen—despite the numbers of condemned British vessels brought into its harbor—but the weather had not been kind. "This poor country, which you once knew in a flourishing state," wrote a French resident of Saint Pierre to a merchant in Bordeaux in June 1758, entrusting the letter to a Dutch vessel that was subsequently seized by the British, "has been afflicted within three years by two dreadful hurricanes, which has for ever ruined one-third of the inhabitants; by a war, more dreadful than can be imagined; and to complete our misery, by a drought, which continued from the beginning of this year, ruins all the plantations, and affords a very melancholy prospect for the ensuing year. May God have mercy upon us, and at least grant us a peace, no matter on what conditions!"[11]

As much as Pitt would have liked to oblige by way of a conquest, he was not without his own limitations. Despite certain successes in North America in 1758, two major raids on the French coast—at Saint-Malo and in Brittany—had ended in failures reminiscent of Rochefort the year before. Many in England were still nervous about the threat of a cross-Channel invasion from France. It was bad enough that Pitt had thrown troops and ships against the French in Canada, but now he wanted to do the same in the Caribbean. Lord Anson, first lord of the Admiralty, opposed diverting more ships from the English Channel. Others remembered Admiral Vernon's less than stunning show in the Caribbean in 1742 during the War of Jenkins' Ear and opposed any expeditionary force in that direction. "Martinique is the general notion," wrote Horace Walpole, but "others now talk of Guadeloupe. . . . It is almost impossible for me to find out the real destination . . . and I would rather not be told what I am sure I shall not approve."[12]

Such criticism aside, Pitt was well past working without a net, and he did not hesitate. Perhaps most importantly, he suddenly found the king in his corner. George II, who might be said to have merely tolerated Pitt in the early years, had suddenly caught the fever of global empire and in at least one critical exchange with Pitt and the duke of Newcastle become one of Pitt's strongest supporters. Praising Pitt's successes to Newcastle, the king lectured the duke that "we must keep Cape Breton [Louisbourg], take Canada, drive the French out of America, and have two armies in Germany, consisting together of 80,000 men." To Pitt, George avowed that "we must conquer Martinique as a set-off to Minorca."[13]

It was nice to have George's blessing for once, but in reality Pitt had already begun phase one of a two-phase plan against the French in the Caribbean. Once again displaying his grasp of the interdependency of global positions, as well as a willing eye toward commercial profit, Pitt had ordered attacks against two French trading stations on the western coast of Africa. Much as William Beckford was promoting a Caribbean excursion, a Quaker merchant named Thomas Cummings had whispered in Pitt's ear about the vulnerability of French positions there and the enormous wealth to be had by seizing them. Pitt promptly gave Cummings a trade monopoly in Senegal and sent him off to Africa with two ships of the line and four support vessels carrying a force of 200 marines. This number was less than formidable, but it was enough. Fort Louis at the mouth of the Senegal River surrendered to the British without a shot in late April 1758.

Buoyed by this success, in mid-June 1758 Pitt authorized a second African expedition against the island of Gorée, 200 miles farther south at the mouth of the Gambia River. Led by Admiral Augustus Keppel—whose North American squadron had rendezvoused with Admiral Boscawen in 1755 in the intercept operation that seized the *Alcide* and the *Lys*—the expedition finally sailed in mid-August. By the end of the year, Keppel had captured Gorée and achieved a threefold success. First, a large quantity of gum senega vital to French silk manufacturers fell into British hands and returned a handsome profit. Second, with the loss of Gorée, French privateers in the eastern Atlantic lost their only secure base of operations on the African coast. Third, there was Pitt's first strike against the Caribbean. These trading stations were the source of upwards of 400 slaves sent annually to sugar planters in the French West Indies. Without the cheap supply of fresh labor, French sugar production on Martinique and Guadeloupe was materially weakened.[14]

Meanwhile, the French were well aware of the overall weakness of Martinique and were still trying every angle to entice the Spanish to come to their assistance. As late as Christmas Day 1758, the French foreign minister Étienne-François de Stainville, duc de Choiseul, urged his ambassador in Madrid to present the case that "to ward off the dangers to Spanish possessions" should Great Britain prevail in the Caribbean, the Spanish court "should order twenty-four war vessels into the anchorages of Martinique."[15] Not surprisingly, Spain declined the invitation.

After a false start in abominable weather, a British fleet of seventy-three ships finally sailed for the Caribbean from Portsmouth on November 12, 1758. The flotilla boasted eight ships of the line, including the ninety-gun *Saint George*; a frigate; four bomb ketches; a hospital ship; and sixty-four transports carrying 6,000 regulars, artillery, and supplies. Crossing the Atlantic without incident, the vessels arrived in Carlisle Bay off the British island of Barbados on January 3, 1759. There, a small squadron already on station in the Caribbean under the command of Commodore John Moore joined them. Moore's command included four additional ships of the line, and Pitt's orders placed him in charge of the combined naval units.

The army was a different matter. Pitt favored yet another of his young colonels to lead the assault: in this instance, John Barrington, the brother of the secretary of war. For all of the king's enthusiasm for the Martinique venture, however, George II balked at Barrington's promotion and instead insisted on Major General Thomas Peregrine Hopson for the command. Hopson was a known commodity who had commanded at Louisbourg before its return to France after the War of the Austrian Succession. He had also served briefly as governor of Nova Scotia and been in charge of the troops dispatched from England to aid Lord Loudoun in his aborted effort to recapture Louisbourg in 1757. Pitt had to settle for appointing Barrington as Hopson's second in command. The good news for Barrington—as he reprised the role

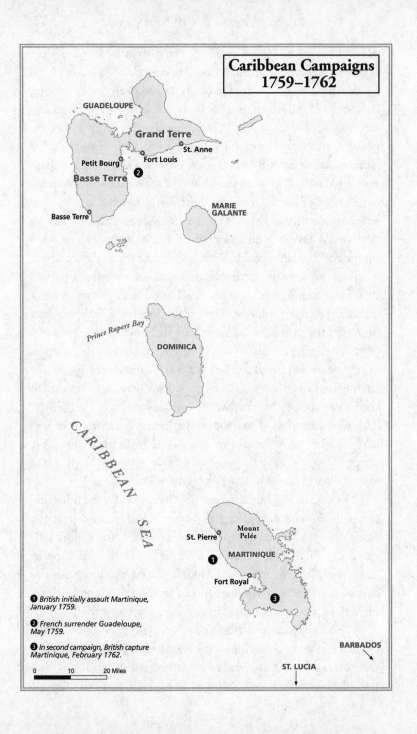

Caribbean Campaigns
1759–1762

GUADELOUPE

Grand Terre

St. Anne

Petit Bourg

Fort Louis

Basse Terre

②

Basse Terre

MARIE GALANTE

Prince Rupert Bay

DOMINICA

CARIBBEAN SEA

St. Pierre

Mount Pelée

MARTINIQUE

①

Fort Royal

③

❶ British initially assault Martinique, January 1759.

❷ French surrender Guadeloupe, May 1759.

❸ In second campaign, British capture Martinique, February 1762.

BARBADOS

ST. LUCIA

0   10   20 Miles

of Lord Howe to Hopson's Abercromby—was that Hopson was a far more decisive commander than Abercromby.[16]

Martinique, with a craggy coastline and deep, rocky ravines, lay about 125 miles northwest of Barbados. Its most dominant feature was 4,800-foot Mount Pelée on the northwestern end of the island. (In 1902, the eruption of this volcano, with a huge cloud of lethal gases, would kill 30,000 people in the nearby town of Saint Pierre.) Martinique was to the windward of Barbados, but its principal towns of Saint Pierre and Fort Royal were on the leeward western side of the island. This meant that the attacking British fleet would have to battle prevailing offshore winds as it hovered off the coast or attempted to enter the inner harbor of Fort Royal.

Hopson determined that Fort Royal, less heavily populated than Saint Pierre, would be his first objective. A system of outer works and artillery batteries flanked the fort itself, which in turn protected the town and the inner harbor. Looking for a sheltered place to disembark his troops, Hopson chose Cas des Navires northwest of Fort Royal and between it and Saint Pierre. Before the landing could be undertaken, Commodore Moore's warships had to silence one artillery battery there and another at nearby Fort Negro. This was accomplished without great effort on the morning of January 16, 1759, and a force of grenadiers and Highlanders under the direct command of Barrington landed at Cas des Navires and marched about a mile to take possession of Fort Negro.

The ease of this initial assault was deceptive. By the following morning, French defenders had begun a steady sniping at the British position from the tangled woods of the surrounding hillsides. The Highlanders made a valiant effort to drive these pesky skirmishers from their protected hiding places, but the terrain was daunting. In the words of one officer, "the Highlands of Scotland, for woods, mountains, canes, and ravines, is nothing to it."

While Hopson and Barrington debated their options, the engineers reported a decided weakness in their position. To reach the road north to Saint Pierre or to march against Fort Royal itself from the landing zone required a move inland of up to six miles via a route that would have to be hacked through this tropical maze. Even if the mechanical effort proved successful, the road would be subject to continuing harassment from French fire for its entire length. Hopson appealed to Commodore Moore to locate a landing site for his artillery closer to the French fort, but Moore replied that none was available. With that news in hand, late on the afternoon of January 17, Hopson ordered his troops back to their transports and abandoned his position. The British invasion of Martinique had lasted but one day.[17]

But Hopson was not easily discouraged. The following morning he convened a council of war onboard Moore's flagship and discussed the possibility of sailing against the prevailing winds and passing right under the guns of Fort Royal to enter the inner harbor. Moore's pilots, who may have voiced their opposition only after an admonishing glance from the commodore, deemed this idea too risky. The consensus was to sail up the western coast some twenty miles and reconnoiter Saint Pierre.

On the northwestern coast of Martinique just south of Mount Pelée, Saint Pierre was guarded by shore batteries and a major fortress just north of the town. Forty merchant ships and assorted privateers lay at anchor under their protection. To test these defenses, Commodore Moore sent Captain Edward Jekyll and the sixty-gun ship of the line *Rippon* to bombard one of the shore batteries. Jekyll easily accomplished this first step, but once again subsequent events proved the toughness of the French defenses.

No sooner had the battery been silenced than *Rippon* was caught in withering cross fire from the fort and other shore installations. To make matters worse, the normally prevailing offshore winds, which had vexed an entry into Fort Royal harbor, suddenly changed to an onshore breeze that pinned *Rippon* against the coast and made maneuvering almost impossible. Finally, Jekyll deployed fifty sailors in longboats to tow the badly battered ship out of harm's way. It was not an experience to inspire confidence. Once again, Moore and Hopson held a council of war.

The Royal Navy could batter down the coastal defenses,

Moore maintained, but his ships would take a beating in the process. If French warships should suddenly appear, his fleet might not be in any condition to repel an attack against Hopson's beachhead—if, that is, Hopson was even able to attain one. The general, too, was having second thoughts about the whole island of Martinique. Even if he was successful in forcing the surrender of Saint Pierre, how could he garrison an island of 30,000 inhabitants who seemed determined to resist with a force one-fifth that number? When Moore suggested that they stand back to sea and head north 100 miles to the town of Basse-Terre on the island of Guadeloupe, Hopson readily agreed.[18]

Guadeloupe was really two major islands separated by a short water passage—mountainous Basse-Terre on the west and flat Grande-Terre to the east. It offered the same strategic attractions as Martinique—major sugar production and a haven for French privateers—but with a smaller population. The capital and major town, Basse-Terre on the western island, was protected by a substantial fort flanked by shore batteries. Its harbor lacked the deep waters of Saint Pierre, and vessels were required to anchor at some distance in the bay.

Whether or not Basse-Terre's defenses were any less formidable than Saint Pierre's is debatable, but early on the morning of January 23, 1759, Commodore Moore dispatched eight ships of the line to bombard the fort and flanking shore batteries. The action was every bit as fierce as what *Rippon* had endured off Saint Pierre, but by late afternoon the French guns fell silent. Apparently, the explanation was not entirely military. "The citadel might have held out indefinitely except that its drinking water cistern was fouled by a mortar shell and the militiamen of the garrison drank rum instead of water all that hot day until disciplinary inhibitions were relaxed and they all went home."

Although there was some thought of landing troops immediately, this order was countermanded and Moore's bomb ketches moved in to soften up the town. The result was an unintended firestorm. The wooden shingles of warehouse after warehouse loaded with sugar and molasses caught fire, and the resulting flames quickly spread to engulf the town. By evening, Moore noted rather sadly in a report to Pitt, "contrary to my wishes," the town became an inferno from "the quantity of rum and sugar that was in it."[19]

The next afternoon, British troops landed without opposition and took possession of the blackened town of Basse-Terre and its fortifications. There was a great debate among the rank and file at the time—and there has been considerable conjecture by historians since—about whether the island of Guadeloupe was indeed the original goal of Hopson's expedition and the dalliance at Martinique merely a diversion. This seems unlikely, but at least one Royal Navy officer avowed from Basse-Terre that the effort against Martinique "was but to feint, for we were intended for this place, which we now have in possession."[20]

But Basse-Terre was only a piece of Guadeloupe, and a full occupation was a long way off. In fact, the British occupied the town of Basse-Terre, but little else—exactly the exposed result that General Hopson had feared if he had been able to take Saint Pierre on Martinique. Finally, three weeks after the landing at Basse-Terre, Commodore Moore ordered a similar naval bombardment against Fort Louis on the southwestern shore of the neighboring island of Grand-Terre. By the time a contingent of Highlanders and marines landed to secure the town, it was in much the same sorry state as Basse-Terre. One difference was that the port at Fort Louis offered the British fleet a much better anchorage.

Two islands, two towns; but the majority of Guadeloupe remained under French control, thanks to some spirited resistance and a little help from the climate. The hot, humid tropics and the fevers they bred were beginning to take their toll on the British troops. By the end of February, over one-fourth of Hopson's troops, some 1,500 men, were sick and had to be evacuated from the island. On February 27, General Hopson himself succumbed to a fever, and command of his troops devolved upon John Barrington.

Barrington was not timid about pushing plans to subdue the remainder of Guadeloupe; but as he and Commodore Moore worked out a plan of operations, they received news that Moore had long dreaded. The French navy had finally made an appearance in this part of the West Indies. Led by Maximin de Bompar, nine ships of the line and three frigates were sighted off Barbados heading for Martinique. Their appearance off Guadeloupe was only a matter of time. If Bompar added the seventy-four-gun *Florissant* from the harbor at Fort Royal to his fleet, and depending on how Moore's ships were scattered along the southern coasts of the two islands of Guadeloupe, the French might prove more than a match for the Royal Navy. Moore decided to assemble the bulk of his fleet in Prince Rupert Bay on the island of Dominica about fifty miles south of Basse-Terre and lie in wait for Bompar. Barrington was going to have to fight on his own.[21]

Barrington acquitted himself well and first moved east on Grande-Terre from Fort Louis to capture the towns of Saint Anne and Saint François. With this flank secure, he turned back to the eastern half of Basse-Terre Island and ground away at the towns of Petit Bourg, Guoyave, and Saint Marie. In the end, only Saint Marie held out, and the French concentrated their troops there in a last stand on the heights above the town. Finally, French resistance melted away after a determined British assault.

On May 1, 1759, the French governor, Nadau d'Etreil, who had repeatedly rejected first Hopson's and then Barrington's entreaties to surrender, was at last forced to do so. More than British guns brought d'Etreil to this decision. With many of the French regulars tied down containing the British garrison at the town of Basse-Terre, French planters on the remainder of the islands had borne the brunt of the coastal fighting. Not only had they grown tired of the conflict by now, but they were anxious to prevent the havoc that Barrington had wreaked on the coastal towns from spreading inland. Pressured by the French inhabitants as much as by Barrington's troops, d'Etreil formally surrendered Basse-Terre and the following day did the same for Grande-Terre. From the British perspective, it was not a moment too soon.

Admiral Bompar's squadron had somehow eluded Commodore

Moore and, while Barrington was occupied against Saint Marie, had landed major French reinforcements at the ruins of Saint Anne. The governor general of the French West Indies, the marquis de Beauharnais, was there in person along with 600 French regulars; 2,000 assorted privateers from Martinique; and 2,000 stands of arms for the local residents. For a moment, it looked as if Barrington was to have a real fight on his hands.

But it was the French planters who came to his rescue. Fearing even more destruction if they repudiated the terms of the surrender, the French planters declined to support Beauharnais's advance, and the governor general was suddenly left without any popular support on the islands. Now, it was Admiral Bompar's turn to fear that an enemy fleet would arrive off his beachhead, and there was little for Beauharnais to do but re-embark his troops and sail back to Martinique.

The main reason that the French on Guadeloupe were eager to avoid continued warfare was that the terms of the surrender had been most generous. First, the British had made a very clear distinction between troops under arms and the inhabitants of the islands. The troops were granted the right to leave Guadeloupe for Dominica or Martinique with full honors of war and with no restrictions on their future military service. In other words, they were not treated as prisoners of war. That alone was unusual. And Guadeloupe's inhabitants fared even better. They were permitted to keep all private property, practice their Catholic religion, educate their children in France, and maintain all the French laws and customs that were then in place. These were to be administered by the existing French authorities, and taxes were to be no higher than those currently paid to Louis XV. There was no requirement to quarter British troops in private houses, and slaves could not be compelled to work on defense fortifications without the consent of their owners.

Essentially, the only changes were that the British Union Jack waved on the flagpole; the island's sugar went to Great Britain instead of France; and molasses was suddenly a legal cargo for Britain's North American colonies instead of smuggled contraband. All this was a far cry from Acadia and the horrors visited there the previous summer. Moreover, some of these terms would have delighted His Majesty's subjects in the British North American colonies, who were beginning to chafe over the quartering of troops and increased taxes.[22]

If they chose to do so, the French could look at Admiral Bompar's naval maneuvers off Guadeloupe and take some satisfaction in the fact that they had still been capable of mounting some sort of an offensive naval operation far remote from France. But such sentiments were hollow. The real victory belonged to William Pitt.

On the two islands of Guadeloupe and nearby Marie-Galante, which the British quickly seized as well, some 40,000 slaves worked on 350 plantations that produced a treasure trove of sugar, molasses, coffee, cocoa, and cotton. Much of it had sat bottled up by the restricted trade of the war, and now French planters—smiling at the most favorable of surrender terms—were eager to ship their produce anywhere it would return a profit. Within a year, 10,000 tons of sugar—more than the amount from all of Great Britain's other holdings in the West Indies—would be exported to Britain in exchange for manufactured goods and more slaves. Within another year, Guadeloupe would be supplying rum distillers in Massachusetts with nearly half of their molasses requirement—three times as much as from Jamaica.[23]

Pitt might still think of Guadeloupe as merely a bargaining chip to be traded for Minorca, but it quickly became much, much more. The coffers of British merchants—the very same middle class that Pitt depended on for taxes to support his far-flung military expeditions—were starting to contain tidy profits. Trade on the high seas, the lifeblood of Pitt's growing empire, was booming, and in the process was starting to sweep the French from the oceans just as surely as any ships of the line. Whatever its original motives, and whatever its intended target, William Pitt's Caribbean gambit had paid off handsomely.

12

Falling Dominoes

Major General Jeffery Amherst and several British regiments arrived in Boston from Louisbourg on September 13, 1758, to much pomp and ceremony. If anything, Abercromby's defeat at Ticonderoga had only heightened the public's desire to embrace the conqueror of Louisbourg, and "the whole town turned out." For his part, the young general was anxious to continue west and see firsthand how bad the situation was north of Albany. Residents of Massachusetts had a different idea. As his regiments camped on Boston Common, their colonial hosts—in a celebratory mood and despite Amherst's wishes to the contrary—"would give them liquor and make the men drunk in spite of all that could be done."[1]

After laying over in Boston for an extra day to let his troops sober up, Amherst marched west across Massachusetts and then northward to Abercromby's camp at Fort Edward. The scene he found was far from encouraging. Abercromby was content to do nothing. Morale among the regular troops was dismal. The provincials and Indian allies were largely gone for the year, the former to bring in what harvest they still could, the latter to return to their villages and hunt for the winter. Any hopes that Amherst held for a strike against Ticonderoga that autumn quickly evaporated. "We have certain advice from Albany, by the courier," the *Boston Gazette* reported on October 16, "that the expedition against Ticonderoga was laid aside for the season; and that General Amherst was upon the road for this place, and is soon expected in."[2]

Indeed, the only good news was that Montcalm and the bulk of his forces had also retired for the winter and posed no threat to Albany. So Amherst sailed down the Hudson to New York City and then went by ship to Boston, hoping to find orders that would permit him to return to England for a winter's respite. None awaited him there, and he sailed on to Halifax to establish winter quarters for some of his troops.

In Halifax on November 9, 1758, Jeffery Amherst received the orders he had been awaiting, but they proved a two-edged sword. The victor of Louisbourg could not return to his beloved England for the winter because Pitt had named him commander in chief of all British forces in North America in place of the plodding Abercromby. Amherst's brother, William, was en route from England with the original dispatches and Pitt's personal assurances that while he regretted denying the young general a return to England, "the King's dependence is entirely upon him to repair the losses we have sustained."

With this news in hand, Amherst elected to return to Boston and then go by road to New York City and establish his winter quarters there. Along the way, he got a sense of how unimpressed some colonials were with his new royal powers as commander in chief. At Branford, Connecticut, he was required to obtain an order from the local justice of the peace to permit his wagon to travel because it was Sunday. "At Norwalk he met Governor [Thomas] Fitch, whose principal interest in meeting Amherst appeared to be to protest against the request to supply wood for a regimental guard stationed there. 'I never heard any petty fogging attorney more equivocating or half so silly on any subject as he was on this,' Amherst noted in his journal, 'but I persuaded him at last to order the delivery of the wood.'"

These experiences to the contrary not withstanding, Amherst did note Connecticut's more redeeming features: "New Haven has a college built with brick and another building of wood that make a fine appearance." Soon, though, Amherst would have much more to worry about than the accoutrements of Yale.[3]

Even as he embarked for Louisbourg the previous spring, Jeffery Amherst had in his possession a secret intelligence report describing much of New France as "ripe for revolt." Its inhabitants, the theory ran, were so tired of war and the continuing famine that many wished for the British to simply deliver them from their "misery." Despite the hardships, however, the Canadians had shown no sign of cracking in the face of the campaigns of 1758, and Amherst now had little reason to think that 1759 would be much different.[4]

That is not to say that New France was not eager for peace. In fact, this was about the only thing that Montcalm and the governor general, Vaudreuil, could be said to agree on. "Without the peace we need, Canada is lost," Montcalm wrote just before he departed Ticonderoga in September 1758. "Peace appears to me an absolute necessity," the governor wrote, quite independently, more than 100 miles away in Montreal the following day.

But peace was something that had to be determined between London and Paris and dictated worldwide—not just along the waterways of the Saint Lawrence. Until those in power took that action, there was nothing for Montcalm and Vaudreuil to do during the winter of 1758–1759 but count the declining rations and debate their diverging views. Montcalm favored incorporating Canadian militia into the regular French units and concentrating Canada's defense along the Saint Lawrence. Vaudreuil was determined to reoccupy Fort Frontenac and reinforce the links to the western posts, particularly Fort Niagara.

The rivalry between the two men was not softened when Bougainville returned to Montreal from France in the spring of 1759 and reported that Montcalm's victory at Fort Carillon had made him the darling of the French court. Vaudreuil's name was conspicuously absent in Louis XV's praise, but the real blow to the governor's prestige came when Bougainville delivered orders elevating Montcalm to commander in chief of all forces in Canada. Henceforth, Vaudreuil was to answer to Montcalm, not the other way around. "I may not look like the man of the hour in Canada," Montcalm boasted in May 1759, "but that is what I look like in Paris."[5]

But the clock would soon toll the fateful hour in Canada, not at Versailles. Unfortunately for both Montcalm and Vaudreuil, the same problem that had long vexed the French perspective in this war still persisted. The situation had always looked different to those ensconced at Versailles and to those charged with defending the wilderness of North America. For his part, Amherst was of a single mind as to his mission. Writing to Pitt about routes of advance for the campaigns of 1759, Great Britain's new commander in chief in North America declared that "by whichsoever avenue we succeed, Quebec or Montreal, Canada must fall, and with it everything on this side of the River Saint Lawrence."[6]

Amherst's letter to Pitt crossed in the mail—if such a phrase can be used to describe the months-long delay in transatlantic communications—with Pitt's own detailed expectations for the campaigns of 1759. Wolfe was off on his own to Quebec, ostensibly with an independent command, but he was expected to coordinate his efforts with Amherst's nonetheless. Amherst was to send both regulars and provincial troops to assist him. In the west, Pitt ordered the rebuilding of Oswego and the pushing of offensive operations at least as far as Fort Niagara. Strengthening Fort Pitt was a given, but all these efforts left Amherst decidedly short of men and matériel to undertake the key campaign Pitt left "to be under your own immediate direction"—the long-awaited reduction of Ticonderoga and Crown Point (Fort Saint Frédéric) and the push northward from there to Montreal.

If Amherst foresaw more glory at Quebec or even Fort Niagara, Pitt tried to soften any blow to his ego by declaring that Amherst's central location on the New York frontier was essential to his duties as commander in chief in cultivating the support of the various colonial governors. (Amherst had already had a taste of that during his first encounter with Connecticut's governor Fitch.) Pitt's concern was that most of the promises for aid and compensation that he had made to the colonials the year before had yet to be paid in pounds sterling. Now, almost as a bothersome afterthought, Pitt added to Amherst's orders that "you will, in case of necessity," draw bills for "any extraordinary expenses incurred for this service."[7]

In other words, just as Forbes had had to be a diplomat to the Indians as well as a military commander the year before, now Amherst was called on to be both of these plus a procurer of funds for the operation of his armies. He would also be required to be a mediator not only between proper British regulars and their less disciplined provincial counterparts, but also among the colonial legislatures. In many respects, it was a thankless task, but he fell to it with decisiveness. His first step was "to pardon any deserter from a regiment in America who shall voluntarily join his colors on or before the 1st day of March." Coming to terms with the colonial legislatures would be harder.[8]

Pitt's government still owed the Pennsylvania legislature 180,000 pounds for supplies and services—including lost wagons and horses—related to Forbes's advance. Only begrudgingly did the assembly advance another 50,000 pounds for the operations of 1759 along the Forbes road, and then only after attaching a string of conditions to the loan. Amherst did a little better in New York, where the assembly approved his request for a loan of 150,000 pounds and issued one-year bills of credit "to enable him to pursue his operations, and facilitate the success of His Majesty's arms."[9]

In early March of 1759 Amherst sent Robert Rogers with ninety of his rangers, 217 regulars from Gage's light infantry, and fifty Mohawk allies to make a scouting expedition to Fort Ticonderoga and map in detail the state of its defenses. Any spring assault against this fortress, however, would have to be tempered by conditions on Forbes's road to Fort Pitt. The garrison that Forbes had left there the previous December had managed to hold the forks, but its supply line to Fort Bedford and Raystown was tenuous at best. As the road dried out in the spring and became passable, French regulars, Canadian militia, and their remaining Indian allies engaged in a series of quick raids all along the 80 miles between Fort Pitt and Fort Bedford. "It is a thousand to one but this letter is intercepted by the enemy," a correspondent wrote on April 17 from Fort Ligonier, smack in the middle, "as the road is waylaid from Pittsburg to Bedford."[10]

Fort Bedford and Raystown were not safe either. "This morning an express arrived here from Raystown," came a report from Winchester six weeks later, "that as thirty of our wagons were coming up with provisions, under an escort of a hundred men, [they] were attacked by 150 French and Indians, who killed and wounded thirty of our people, destroyed all the wagons, and carried off most of the provisions on their horses." It was no wonder that the *Maryland Gazette* of Annapolis reported on June 14 that "we cannot find that the report lately spread of the French having retaken Pittsburg has any truth in it; but have the strongest reason to hope that it was altogether without foundation."[11]

The dubiousness of the report was certainly not because the French hadn't tried to recapture Fort Pitt. After blowing up Fort Duquesne, Commandant Lignery had retreated only as far as Fort Machault at Venango, about 100 miles up the Allegheny from the forks, and spent the winter there. While French elements were harassing Forbes's road the following spring, Lignery was busy gathering forces at Venango for what he hoped would be a lightning descent of the Allegheny to recapture Pitt before spring reinforcements could reach it.

Among those French troops assembling at Venango were units from the garrison at Fort Niagara. Thinking Niagara well to the rear of any frontline action—and evidently discounting any repeat of Bradstreet's foray against Fort Frontenac the previous summer—Commandant Pierre Pouchet had been only too glad to provide support for Lignery's counteroffensive. Meanwhile, of course, in part to take pressure off this anticipated threat to Fort Pitt and to secure Forbes's road, Amherst acted quickly and quietly to undertake Pitt's directed strike west against Fort Niagara itself.

To lead this western incursion, Amherst chose John Prideaux, who had become colonel of the Fifty-fifth Regiment on the death of Lord Howe and who now with this new command assumed the rank of brigadier general in North America. If Prideaux needed any recommendation, there was none better than Lord Ligonier's assertion to Amherst that he was "a very active, diligent officer." He was also discreet. As Prideaux assembled some 3,000 regulars, including the Forty-fourth and Forty-sixth regiments, a battalion of Royal Americans, and a company of Royal Artillery, and led them west up the Mohawk, secrecy was Amherst's sternest charge. "I kept my intended operations secret," wrote the commander in chief. "If the Indians know them the French will have it."[12]

Indians, however, were to be an important part of Prideaux's force—thanks in some measure to Sir William Johnson. Part of Pitt's charge to Amherst as commander in chief had been to recruit Indian allies to the British cause and counter their long use by the French. Now, that finally appeared to be happening. Although Johnson was always inclined to overstate his personal value in a situation, he now reported that he had been "able to prevail upon the greater part if not the whole of [the Iroquois] to join His Majesty's Arms."[13]

The sudden inclusion of 1,000 Iroquois warriors in a British command en route to attack the French on their own ground showed just how much had changed in the year since the Iroquois had failed to support Bradstreet's attack against Fort Frontenac. In actual fact, this turnaround had very little to do with Johnson's sway over the Iroquois. He held little unless it suited their purposes, and now it did.

Leery of the increasing flow into their territory of western Indians from beyond the Great Lakes disgusted by the inability of the French to supply high-quality trade goods at competitive prices, and impressed by the recent capture of Fort Duquesne, the Iroquois at last allied themselves with the British. Or so they said. A year before, Sir William Johnson had the goal of maintaining the Iroquois's neutrality and preventing them from joining the French. Now, he was leading them against Fort Niagara, which was held by the French.[14]

Brigadier Prideaux's force moved up the Mohawk and rendezvoused with Johnson's Iroquois at the ruins of Oswego on June 27, 1759. Prideaux delegated Colonel Frederick Haldimand to remain at Oswego with half of the Royal Americans and over 500 New York provincials and begin the task of rebuilding Fort Oswego, effectively adopting Forbes's strategy of establishing strong outposts on his line of advance. Then, with 2,000 troops and 1,000 Iroquois, his force rowed west in whaleboats and bateaux along the southern shore of Lake Ontario. On July 6, they landed unopposed at the mouth of the Little Swamp River about three miles to the east of Fort Niagara and the mouth of the Niagara River. The surprise was complete. Commandant Pouchot, who had dispatched troops to aid Lignery's effort to recapture Fort Duquesne, suddenly had his own problems.

First occupied by the French in 1725, Fort Niagara had been greatly strengthened during the course of the war and was in a much better state of military preparedness than Fort Frontenac had been the year before. It sat on a rocky promontory of land just east of the mouth of the Niagara River, with Lake Ontario to the north and the river curving around its western and southern sides. The fort's principal defense was a 900-foot-long battlement anchored by three bastions extending along its entire eastern side. Other than the requisite barracks, armories, and support buildings, its principal interior structure was a three-story stone building that served as the commandant's headquarters and was called the House of Peace, a designation deriving from its role as an Indian trading center. Outside the walls, there was a dock on the Niagara River, and there were various earthworks to the east

If there were weaknesses at Fort Niagara, they were twofold. First, as was usually the case by now, the French were lacking in troops. Having sent troops to Lignery, Pouchot had left fewer than 500 soldiers in his garrison: by one account 150 French regulars, twenty-three artillerymen, 180 colonial marines, and 133 Canadian militia, plus a few remaining Indian allies. The second weakness was that most of Niagara's defenses presumed an attack from the east. The opposite promontory on the western side of the mouth of the Niagara River, Montreal Point, was undefended and offered a powerful artillery position should it be occupied by an attacker.[15]

Accordingly, as Prideaux landed his forces at the mouth of the Little Swamp River, he immediately ordered men to portage whaleboats loaded with three howitzers and ammunition through the woods and into a ravine south of the fort. Called La Belle Famille, the ravine led down to the Niagara River. Dodging French artillery fire, this little flotilla ferried the howitzers across the Niagara to its western bank early on the morning of July 7 and proceeded to erect an artillery battery opposite the fort.

The following day, having moved his forces to within 1,000 yards of Fort Niagara's eastern side and begun to dig siege trenches, Prideaux sent a flag of truce to Pouchot and demanded his surrender. Pouchot made no reply, but he had already frantically sent messengers to warn a French outpost at the main portage around Niagara Falls and—more important—to beg Lignery to call off the expedition against Fort Pitt and rush to his aid.[16]

As Pouchot waited expectantly for some sign of relief, Prideaux's men dug siege trenches closer and closer to the walls of Fort Niagara. Meanwhile, there now began a series of parleys between the 100 or so Seneca who were still on the French side inside Fort Niagara and the Iroquois who were among the attackers outside the fort. Kaendaé of the Niagara Seneca visited with Johnson's Iroquois and tried to persuade them to withdraw together up the Niagara and leave the French and the British to fight between themselves. This plea to return to the time-honored neutrality that the Iroquois Confederation had long practiced failed only after Johnson promised "his" Iroquois "the first chance to plunder the fort after it fell."[17]

While these negotiations dragged on, Pouchot acquiesced in the accompanying delays because they bought him time for Lignery's hoped-for relief column to draw nearer. Prideaux was also content to tolerate the parleys, because he used the time to dig trenches even closer. Finally, Kaendaé's Seneca were permitted to withdraw from Fort Niagara under a flag of truce. Pouchot may have realized that they were now reluctant warriors and was probably relieved to see them go. These Seneca, however, repaid their French hosts by attacking the French outpost at the Niagara Falls portage where Pouchot had sent his oxen and cows to keep them from falling into British hands. After butchering the livestock, the Seneca "actually carried the meat to the English camp."

Meanwhile, Johnson's Iroquois were in contact with the Indians accompanying Lignery's approaching relief force. These Indians were doing their own parleying to dissuade the Iroquois from further participation in the campaign. The Iroquois permitted messages to Pouchot to get through the siege lines and in retrospect appear to have promised to sit this one out if the advancing Indians did the same.[18]

On July 17, the three British howitzers hauled to Montreal Point began shelling Fort Niagara's dock and raining fire upon the House of Peace and other interior buildings. Two days later, one British battery was within eighty yards of the eastern wall. One account maintains that the French had fired 6,000 cannonballs at the attackers so far but managed to kill only three men and wound twenty more. The number seems highly inflated, but it may indeed be true because General Prideaux offered a bounty of sixpence in New York currency for every twelve- and nine-pound shot his troops recovered in good enough condition to be fired back.

But then on the evening of July 20, General Prideaux and his aide, Colonel John Johnstone, were making an inspection of the front trenches when a French musket ball killed the colonel on the spot. Prideaux helped remove Johnstone's body to the rear and then resumed his own inspection at the front. The general had barely returned when he stopped to watch a crew fire a newly installed mortar. The artillery piece blew up with a roar and a hail of heavy shrapnel. A chunk of the exploding barrel blew away a sizable piece of Prideaux's skull, killing him instantly.[19]

Now there arose a question of command. The provincial colonel Sir William Johnson was only too happy to oblige, but Lieutenant

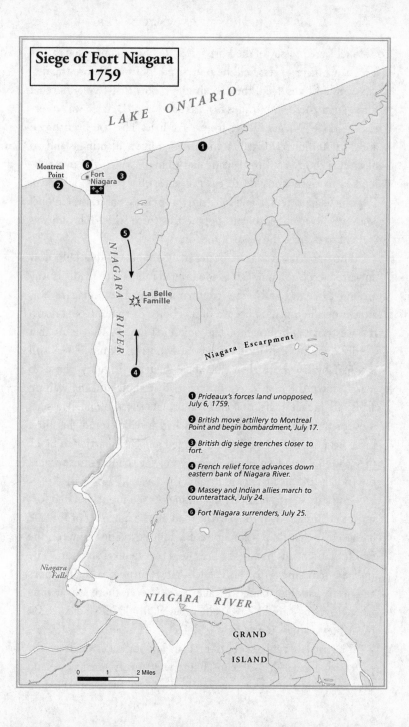

Siege of Fort Niagara 1759

LAKE ONTARIO

❶

Montreal Point
❻
❷ ⚑ Fort Niagara ❸

❺

N I A G A R A   R I V E R

✷ La Belle Famille

❹

Niagara Escarpment

❶ Prideaux's forces land unopposed, July 6, 1759.

❷ British move artillery to Montreal Point and begin bombardment, July 17.

❸ British dig siege trenches closer to fort.

❹ French relief force advances down eastern bank of Niagara River.

❺ Massey and Indian allies march to counterattack, July 24.

❻ Fort Niagara surrenders, July 25.

Niagara Falls

N I A G A R A   R I V E R

GRAND

ISLAND

0   1   2 Miles

Colonel Eyre Massey of the Forty-sixth Regiment contended that as a regular officer in the field he outranked Johnson. Before the dispute could be resolved, they received a report that Lignery's relief force, augmented with troops from as far away as Detroit, was entering the upper Niagara River from Lake Erie. The 600 Frenchmen and a 1,000 Indians looked to one observer like a "floating island, so black was the river with bateaux and canoes." What the two contenders to command said to each other at this point is unknown. Johnson's supporters would long sing his praises, but it was Colonel Massey who marched out to oppose this force while Johnson was content to oversee the siege of the fort.[20]

The good news for the besieged Pouchot was that help now appeared to be at hand. But as the French relief force made its way around Niagara Falls, the commandant at Fort Niagara was almost beside himself because of its choice of route. Come down the western bank, Pouchot had advised Lignery. Rid us of that deadly redoubt of British artillery at Montreal Point. It was still manned by no more than 200 men. Instead, Lignery chose to march down the east bank portage road straight into the waiting arms of a picket guard of some 100 provincials. Colonel Massey's movement with about 450 regulars had just bolstered this little force near La Belle Famille and erected an abatis barrier to block the road. Six hundred Iroquois waited in the shadows to see what the Indians of the advancing French force would do.

The answer was not long in coming. As the French moved forward in column down the narrow road on the morning of July 24, it was clear that they were alone. Further entreaties between the two Indian groups had indeed resulted in a neutral standoff. The Indians marching with the French had returned to their canoes and were paddling back to Lake Erie. If ever there was a time when Lignery should have remembered Braddock's advance on the Monongahela, it was now.

The mixed French force of regulars and militia came down the road with a great deal of élan. There was "a very great noise and shouting" and some troops began to discharge their muskets into the abatis ahead. For their part, the British troops stood silent behind their cover and waited until Lignery had unwittingly led the French tight inside the trap. When the French were within thirty yards of the abatis that now flanked them on either side of the road, Colonel Massey gave the order to fire. "The men," Massey later reported, "received the enemy with vast resolution, and never fired one shot until we could almost reach them with our bayonets."

The French tried desperately to deploy from a long column into line formations, but it was indeed the Monongahela all over again, though with opposite roles. The British poured seven deadly volleys at almost point-blank range into the French formation. Only when the French troops at last gave way under a British bayonet charge did the Iroquois join the fray and pursue their retreat. The result was unspeakable carnage.

Only about 100 of the French force—mostly the wounded and a handful of officers—survived to be taken prisoner. Massey did not restrain himself when he later wrote an account of the engagement to Pitt. "As I hear the Indians have got great credit by that day, in Europe," wrote the colonel, "I think I would not do justice to the Regiment [the Forty-sixth] . . . if I would allow savages, who behaved dastardly, to take that honour, which is deservedly due, to such of His Majesty's troops, as were in that action."

Able to hear the sounds of the battle, but unsure of its results, Commandant Pouchot at Fort Niagara continued his daily artillery battle with the British positions. Now, he refused to believe either the report of a loyal Onondaga or that of a British officer who appeared before his gates with a white flag. Only after one of his own officers was conducted into the wounded Lignery's presence in the British camp did Pouchot accept the inevitable. The following day he surrendered Fort Niagara to Sir William Johnson.

Despite his demands to march from the fort with full honors of war and be allowed to conduct his men to Montreal, Pouchot and his troops were instead sent to Albany as prisoners of war. The only saving grace was that Sir William Johnson managed to keep the Iroquois in check with only the material plunder of the fort.

There would not be a reprisal for Fort William Henry. For Lignery's part, he did not live long enough to suffer humiliation or to see the fleur-de-lis disappear from the Ohio Valley forever. [21]

Meanwhile, General Amherst was trying to make his own progress against the French fortresses of Carillon (Ticonderoga) and Saint Frédéric (Crown Point). He was not a plodder like Abercromby, but neither was he one to dash off without thorough plans. In addition to the slow arrivals of provincial troops, one nagging problem was still a lack of provisions and a lack of adequate funding from the colonial legislatures. In late April, Pennsylvania came through with another loan—this one of 100,000 pounds—but Amherst was nonetheless reduced to pleas to the local populace to supply his troops with sustenance.

"Whereas the army destined for Lake George and beyond it, cannot be too well supplied with all kinds of refreshments, which must greatly contribute towards the preservation of the health of the troops," Amherst's circular read, "these are therefore to make known to the people of the province, that such of them who shall be inclined to convey to the lake, live cattle, sheep, fouls, eggs, butter and cheese or any other refreshments or necessaries whatever (rum and spirituous liquors excepted),* . . . shall meet with all protection."[22]

Now in June 1759, having sent support to Wolfe en route to

---

* The general's prohibition of rum and spirits did not extend to spruce beer, which was a highly recommended and widely used remedy for scurvy. Amherst himself took note of the recipe in his journal: "Take 7 Pounds of good Spruce & boil it well till the bark peels off, then take the Spruce out & put three Gallons of Molasses to the Liquor & boil it again, scum it well as it boils, then take it out the Kettle & put it unto a cooler, boil the Remainder of the Water sufficient for a Barrel of thirty Gallons, if the Kettle is not large enough to boil it together, when milk warm in the cooler put a pint of yeast into it and mix it well, then put it in the Barrel, and let it work for two or three days, keep filling it up as it works out. When done working, bung it up with a Vent Peg in the Barrel, to give it Vent every now & then, it may be used in two or three days after. If wanted to be bottled it should stand a fortnight in the cask. It will keep a great while." (J. C. Long, *Lord Jeffery Amherst: A Soldier of the King* [New York: Macmillan, 1933], p. 110.)

Quebec and dispatched the efforts to reinforce Fort Pitt and reduce Fort Niagara, Amherst assembled his own forces on Lake George, which—in Parkman's words—"for five years past had been the annual mustering-place of armies." Here, Amherst paused to begin construction on a replacement for Fort William Henry that was to be called Fort George. By the time his army finally embarked in bateaux and rowed north down Lake George, it was July 21. This time, the advancing army numbered about 10,000 men, considerably less than Abercromby's force. The slaughter of the previous year, however, meant that the troops advanced with far greater trepidation.[23]

But the result this year was to be decidedly anticlimactic. On the morning of July 23, as Amherst surveyed the French defenses above the field that Abercromby had never personally seen, the general detected a flurry of movement in the French trenches. It appeared that the French were pulling back. But to where? Into the fort, down Lake Champlain, or on some mission to outflank him?

Cautiously, Amherst avoided the open field in front of the defenses and ordered an advance along the thickly wooded lakeshore. As they reached the French entrenchment, the lead troops looked up to the stone walls of the fortress 800 yards away and expected to be met with a hail of artillery and musket fire. Instead, they were greeted only by silence. Then, their eyes were drawn down to the level of the abandoned breastworks. In the center of the trenches, the French had dug a large grave and placed a cross with an engraved copper plate at its head. "Bury their generals here, like Oreb and Zeeb, Zebah and Zalmunna," it read, in a biblical reference to the enemies struck down by Gideon.[24]

The French army of some 3,000 under the command of Brigadier General François-Charles de Bourlamaque had retreated north to Fort Saint Frédéric, leaving only a garrison of 400 troops to hold Fort Carillon. The British occupied the French trenches, brought up six twenty-four-pound artillery pieces, and wondered what might happen next. The answer came with a gigantic explosion just before midnight on the evening of July 26. The French garrison had slipped away in boats down Lake Champlain and left a fuse burning to the fort's well-stocked powder magazine. The conqueror of Louisbourg had become the conqueror of Carillon with the loss of only five dead and thirty-one wounded.

If not a plodder, Amherst was definitely a fort builder, and he set to work reconstructing the Carillon ruins as Fort Ticonderoga. If he was somewhat of a teacher on this expedition, Amherst had with him a host of willing pupils from among the colonial troops. Many of their names would be heard again along the waters of Lake Champlain.

With Amherst at Ticonderoga that summer—in addition, of course, to Robert Rogers—were Israel Putnam of Connecticut, Philip Schuyler of New York, and John Stark of Vermont. Also present was Ethan Allen of Vermont, who would pay his own nighttime call upon a British garrison at Ticonderoga sixteen years hence and demand its surrender "in the name of the Great Jehovah and the Continental Congress." Finally, there was a young lad from Connecticut who would stand beside Allen on that day and march on to Quebec in 1775. His name was Benedict Arnold.

With Carillon in hand, the next objective to be considered was Fort Saint Frédéric, but Amherst barely had time to send a scouting party of rangers to look at its situation before they reported back that the French had abandoned it, too, and also detonated its powder magazine. Arriving at Fort Saint Frédéric on the evening of August 4, Amherst looked toward the rise immediately west of the French ruins and pronounced it "the finest situation, I think, that I have seen in America." Here, he would build Crown Point, a position that "will have all the advantages of the lake . . . that can be wished for."[25]

As he surveyed this ground at Crown Point that evening, Jeffery Amherst received a message from Sir William Johnson announcing the capture of Fort Niagara. The French had attacked Oswego in Johnson's absence, but Colonel Haldimand had beaten off the assault and sent those French troops scurrying for the Saint

Lawrence. Meanwhile, Brigadier John Stanwix was marching north from Fort Pitt to occupy the French forts of Venango, Le Boeuf, and Presque Isle. The few French survivors who escaped the carnage at La Belle Famille fled west to Detroit and abandoned all claim to these Allegheny posts. Suddenly, it appeared that the fall of Fort Niagara had opened a vast void in New France from the forks of the Ohio almost to the gates of Montreal. Now, with the capture of Carillon and Saint Frédéric, Amherst was eyeing those very gates.

Venango, Le Boeuf, Presque Isle, Niagara, Carillon, and Saint Frédéric—they had all fallen like dominoes, far more easily than might have been surmised at the beginning of the summer. It only remained for Amherst to learn how Brigadier James Wolfe was faring in front of the biggest domino of them all—Quebec.

13

BATTLE FOR A CONTINENT—
OR IS IT?

We left Brigadier James Wolfe in August 1758, as he was savoring the fall of Louisbourg and champing at the bit to sail up the Saint Lawrence River to attack Quebec. When Admiral Boscawen and General Amherst deemed the season too late and the news of Abercromby's defeat at Ticonderoga too alarming to let them concur, Wolfe was instead dispatched with three regiments to burn French fishing villages on the Gulf of Saint Lawrence. Observing in a letter to his father that his mission was "to rob the fishermen of their nets, and to burn their huts," Wolfe reported to Amherst after the fact that "we have done a great deal of mischief—spread the terror of His Majesty's arms through the whole gulf; but have added nothing to the reputation of them." Then, without specific orders to do so, Wolfe left his officers and men at Louisbourg and abruptly sailed for England onboard Admiral Boscawen's flagship.[1]

No one was more surprised to learn of his arrival there than William Pitt. After the fall of Louisbourg, Pitt dispatched orders that Wolfe was to remain in North America for the winter and prepare for the spring campaign—wherever that might be. Now, here was the young brigadier suddenly making the rounds between London society and his old regiment and, in Amherst's absence, being hailed as the man of the hour for the capture of

Louisbourg. Never mind that Wolfe had left there before Pitt's orders to stay arrived—again, transatlantic communication was slow. The part most disconcerting to Pitt was that Wolfe had left Louisbourg without *any* orders.[2]

When military friends alerted Wolfe to his faux pas, the briefly humbled man of the hour wrote a short but ingratiating letter to Pitt alluding to his understanding that Lord Ligonier had previously agreed to Wolfe's return to England after but one campaign season. Amherst, too, of course, had received similar assurances from Ligonier, but he had dutifully circled between New York, Boston, and Halifax waiting for the orders that deprived him of a winter's respite in England. Now enjoying just that, Wolfe assured Pitt that he had "no objection to serving in America, and particularly in the river Saint Lawrence, if any operations are to be carried on there."

If Wolfe was being coy in that last phrase, Pitt overlooked it, as well as his indiscretion in returning. Of course, there would be a campaign up the Saint Lawrence against Quebec, and for better or worse, Pitt had concluded that the soon to be thirty-two-year-old brigadier with the shock of flaming red hair to match his fiery ego was the man to command it.[3]

But Pitt kept Wolfe on the hook just a little while longer. "I am going directly to the Bath to refit for another campaign," Wolfe wrote to a military comrade in reference to England's most famous spa. If he knew where that campaign might be, Wolfe gave no indication, but he was not without his usual criticisms. "We shall look, I imagine, at the famous post at Ticonderoga," Wolfe continued, "where Mr. Abercromby, by a little soldiership and a little patience, might, I think, have put an end to the war in America."[4]

Soldiering, however, was not all that was on Wolfe's mind at Bath. He renewed an acquaintance that he had made the previous year before sailing for Louisbourg. Her name was Katherine Lowther, and she was the daughter of a former governor of Barbados. Ten years before, Wolfe's courtship of another woman had fallen flat, in part because of his absences in military service.

"Young flames," he confessed at the time, "must be constantly fed, or they'll evaporate."

Now, the matter seemed to take on far greater urgency for Wolfe at thirty-two than it had when he was twenty-two, and he pressed it with great vigor. What Miss Lowther saw in the gangly, thin, frequently fussy brigadier is open to conjecture, but his standing had certainly been raised over that of the previous year by his exploits at Louisbourg. For his part, Wolfe seems to have been transfixed by her captivating eyes. In the span of two months that were to be increasingly busy with military plans, Wolfe somehow found the time and the ardor to court her. He would leave England with both Katherine's miniature portrait and her promise of marriage on his return.[5]

Just before Christmas 1758, Wolfe's courtship of Katherine Lowther was interrupted by a summons to meet with Pitt at his country home. There, Wolfe finally received his orders for Quebec along with the temporary rank of major general in North America. Far from being humbled by the appointment, Wolfe got himself back into hot water almost immediately by adamantly insisting to Lord Ligonier that he had the right to name his own brigadiers. Ligonier acquiesced in two of Wolfe's choices, but held firm with regard to George Townshend as the third, despite Wolfe's threats. No wonder Wolfe was described as "intensely human, subject to error, not without vainglory, quick of temper, sanguine, emotional, vehement to a fault"—and this from the pen of a generally admiring biographer. To put it more succinctly, Wolfe was an odd duck.

To this point, there is an anecdote about Wolfe that must be told. Its details are perhaps dubious—or at least embellished—but its overall tenor is such that this same biographer proclaimed that it "by no means deserves to be rejected *in toto*." On the eve of his departure for North America, Wolfe was invited to dine with Pitt and Lord Temple, Pitt's brother-in-law, to receive, it is assumed, Pitt's last-minute admonishments for the conduct of the coming campaign. As the evening progressed, Wolfe's tongue was loosened more than normally by the wine. (It appears that as a rule he drank sparingly.) Not only did Wolfe freely offer his own senti-ments on the task at hand, but "he drew his sword, he rapped the table with it, he flourished it round the room, [and] he talked of the mighty things which that sword was to achieve."

Pitt and Temple sat speechless at the display. If Pitt had second thoughts about his chosen commander, he confided them only to Temple after Wolfe had departed. It was Temple who whispered of Wolfe's conduct to the duke of Newcastle, who was still Pitt's part-ner in government. Newcastle is supposed to have hurried to tell George II of Wolfe's outburst as further evidence that Pitt's cadre of young brigadiers was not to be trusted with the fate of the empire and that Wolfe in particular was quite insane. "Mad, is he?" retorted George, perhaps humbled by his insistence on the defeated Abercromby and the humiliation of his own son in Europe. "Then I hope he will bite some of my other generals!"[6]

We can take that with a grain of salt, but here is what hap-pened next. Wolfe sailed from Portsmouth on February 14, 1759, onboard *Neptune*, the flagship of Vice Admiral Charles Saunders. In England, he had been promised some 12,000 troops, but by the time his army embarked for Quebec from Louisbourg, the num-ber was closer to 9,000. All were regulars or Royal Navy marines with the exception of six newly raised ranger companies. Always a critic of the colonials, Wolfe labeled these North Americans "the worst soldiers in the universe."[7]

After the dust had settled with Ligonier, Wolfe's brigadiers were Robert Monckton, James Murray, and George Townshend. Monck-ton was the most experienced in North American affairs, having accomplished the only British success of 1755 with his capture of Nova Scotia's Fort Beauséjour. Murray was perhaps an odd choice. Wolfe had had a run-in with him in Scotland years before, but after watching his service at Louisbourg as a regimental commander, Wolfe had praised him with a rare accolade, noting that "the public is much indebted to him for great service in advancing . . . this siege."[8]

Townshend was more political than military in nature, but he had cut his teeth alongside Wolfe and Monckton at the battle of Dettingen so long ago. A falling-out with the duke of Cumberland left him out of the army for a time, but now Townshend seemed determined that some of Wolfe's glory should rub off on himself. As both the nephew of the duke of Newcastle and a political ally of William Pitt in Parliament, Townshend was in easy touch with both sides of the Newcastle-Pitt marriage and could be counted on to report back to London any warts that might develop on Wolfe's campaign. Of this quartet of Wolfe and his three brigadiers, Wolfe was the youngest by six months. In the end, all three brigadiers would come to detest him. And so, to Quebec they all sailed.[9]

Founded by Champlain as a permanent settlement in 1608, Quebec was the gateway to the heart of New France. The town's very name told its geography. "Quebec" derived from an Algonquin word meaning "the river narrows here." Indeed, it does. For 350 miles inland from the western tip of the Ile d'Anticosti, the Saint Lawrence estuary is fifteen, twenty, even thirty miles wide. Then, just below Quebec, faced by the plug of the Ile d'Orleans, it narrows upstream of the island to barely three-quarters of a mile.

The town stood on the northern shore atop a rocky promontory cut by the Saint Lawrence to the south and its tributary, the Saint Charles River, to the north. The promontory came to a point on its eastern side above the confluence of the two rivers and the waters of the Orleans Basin. Only to the west did the promontory widen beyond the city walls, rise to a shallow ridge, and then fall away gently across what was called the Plains of Abraham— not because of any biblical reference, but because one of Champlain's river pilots, Abraham Martin, had settled there.

Quebec itself was divided into the Lower Town just above the river on a long, narrow strip that was crowded with docks, warehouses, and houses; and the Upper Town astride the tip of the promontory 300 feet above. The Upper Town was completely encircled by a stout wall, and its artillery commanded a wide sweep of the rivers below and plain to the west. On three sides the escarpment fell away from the walls in steep cliffs, while six bastions anchored the wall on its wide western side. No wonder Count Frontenac, when viewing the heights in 1672, exclaimed, "I never saw anything more superb than the position of this town. It could not be better situated as the future capital of a great empire!"[10]

Northeast of Quebec—the Saint Lawrence River maintains its generally southwest-to-northeast flow through here—there was a broad plain along the north bank of the river that extended three miles from the mouth of the Saint Charles to the little village of Beauport. Downstream from Beauport, the riverbank began to rise again the farther east one went until the cliffs crested at the 300-foot falls of the Montmorency River. West of Quebec—above the town in river parlance—the line of cliffs continued on the north shore with only a narrow break at a place called Anse au Foulon.

Stout though its position might appear geographically, Quebec in the spring of 1759 was in a sorry state. In many respects, it mirrored the rest of New France. "The harvest of 1758 had been the worst of the whole war in Canada, and the winter of 1758–59 the coldest in memory." Once again, Quebec and much of Canada were starving. Without provisions from France, it would be difficult to mount a defense against any of the various avenues by which the British appeared certain to threaten.[11]

"If your winter proved melancholy, ours I assure you, has not been gay," wrote one resident of Quebec to his brother-in-law in late April 1759. "We have been reduced to a quarter of a pound of bread per day, and have often had no meat at all. . . . In a word, every thing is above price: 'Tis a most melancholy circumstance to think what will become of us. It is my opinion, if peace does not immediately take place, or great succor sent us from France, our colonies had better submit to the English than continue in their present misery." Now, there was little else to do but wait for the attack that all of Quebec knew was almost certain to occur.[12]

English attacks on Quebec were nothing new. In fact, the English had captured what was then a tiny settlement in 1629, only to have the town and surrounding country returned to France by treaty. Other attempts by the English to capture Quebec in 1690 and 1711 had failed, as had a more recent attempt in 1746. In each case the English had planned an assault up the Saint Lawrence and encountered varying degrees of calamity on it. In fact, the river was a gigantic bogeyman in legend, if not in fact. The very thought of sailing up the Saint Lawrence, with its changing tides and currents, unpredictable shallows and sandbars, and blinding fog made most Royal Navy captains shudder. "I believe the difficulties supposed to attend the navigation of the River Saint Lawrence to be more imaginary than real," wrote Amherst to Pitt, but then the only waterway before Amherst was Lake George.[13]

The problem of the Saint Lawrence belonged to Vice Admiral Charles Saunders. His most recent assignment had been the blockade off Brest and his patron, First Lord of the Admiralty Anson, had tapped him for this important task. When the main British fleet at last weighed anchor and struggled forth in a light breeze from Louisbourg harbor early in June, it numbered 162 ships, including twenty-one ships of the line, five frigates, fourteen sloops, two bomb vessels, a cutter, and 119 transports loaded with Wolfe's men and matériel. There had, however, already been one major naval setback.[14]

When Admiral Boscawen returned to England late in the fall of 1758 with the bulk of the British fleet, he ordered Rear Admiral Philip Durell to remain at Halifax for the winter with some fourteen ships of the line and frigates. Durell was to be prepared for offensive operations early the following spring. In December, Pitt wrote to Durell about the coming campaign up the Saint Lawrence and specifically ordered him to take up station off the mouth of the river in ample time to prevent reinforcements and supplies from France from reaching Quebec.

But when Saunders and Wolfe sailed into Halifax on April 30, Durell's squadron was still in the harbor, held there by late ice floes in the Gulf of Saint Lawrence that were the result of the unusually bitter winter. (Wolfe, who had not been impressed with Durell's actions off Louisbourg the summer before, thought that too easy an excuse.) By the time Durell's ships finally reached the Saint Lawrence, two French frigates and a convoy of fourteen supply ships had managed to slip up the river and reach Quebec.[15]

The effect on the town was electrifying. Not only would its citizens have food, but Montcalm would also have provisions with which to mount a strong defense. Had Durell succeeded in intercepting this relief convoy, the general populace of Quebec might have had little choice but to open its gates at Wolfe's arrival in order to avoid starvation.

So, with Durell's squadron leading the way, Saunders's main fleet sailed up the Saint Lawrence. It helped immensely that the navigator of one of the lead ships, *Pembroke*, was none other than a young lieutenant named James Cook, then still cutting his teeth but in time to become one of the Royal Navy's most distinguished sailors. Between some captured Canadian river pilots, who were lured aboard the first British ships by a false display of French colors, and a bit of verve on the part of several captains, the British fleet passed upriver without the loss of a single ship and anchored off the Ile d'Orleans. Seeing such audacious seamanship, the French themselves were amazed. "The enemy," reported Governor Vaudreuil, "have passed sixty ships of war where we durst not risk a vessel of a hundred tons by night and day."[16]

Less than five miles from his objective, Wolfe now made his opening moves in rapid succession. Troops landed on the Ile d'Orleans on June 26. Three days later, Monckton's brigade went ashore at Point Lévis and soon seized the heights directly across the river from Quebec. By July 12 a battery consisting of four mortars and six thirty-two-pounders was placed on the heights within 900 yards of Quebec. "We opened our battery on the town," wrote one British soldier, "which played its part very well, and soon set several houses on fire, which burned to the ground. The enemy returned the compliment as they could, but did us but little damage." Two additional batteries were soon in place on Point Lévis. Meanwhile, elements of Murray's and Townshend's regiments had landed on the north bank downstream from Montmorency Falls.[17]

But Quebec was far from trembling. The British artillery fire was demoralizing to its citizens, but for the moment their bellies were full and they were content to wait behind its walls. Montcalm, commanding a force of some 14,000 French regulars, marines, Canadian militia, and a handful of Indians, was also content to wait. Rather than concentrate around a single stronghold such as the Upper Town, Montcalm had deployed his forces all along the wide Beauport shore where Wolfe had first proposed to land. His left flank was anchored more than five miles from Quebec by fortifications on the Montmorency River. West of the city, his right flank was protected by militia units stretching upriver another five miles to Sillery.

By spreading out his forces, Montcalm, the man who had successfully besieged Fort William Henry and held firm at Carillon, was determined not to get caught in a trap, but to leave himself with routes of retreat upriver with the bulk of his forces should Quebec fall. It required some communication—and perhaps a little intuition—to know where to mass his troops along this wide front in the event of a concerted attack, but let Wolfe come. Montcalm was content to wait.

James Wolfe had other ideas. The short northern summer was slipping away. Louisbourg had surrendered the year before on August 1, yet Quebec showed no signs of so speedy a fall. On July 31, determined to bring Montcalm into a full-blown engagement, Wolfe ordered a combined amphibious assault from the Saint Lawrence and across the Montmorency River against Montcalm's left flank. It was a disaster.

Warned in advance by the thunder of a British naval bombardment, Montcalm calmly shifted troops to his left. A low tide was necessary for the land-based troops to cross the Montmorency below the falls and attack west, but the low tide also played havoc with the boats of the amphibious landing. Rather than massing on the beachhead—hotly under fire though it was—companies of disembarking grenadiers began a frantic attack inland, perhaps remembering the impetuousness of those Highlanders who had

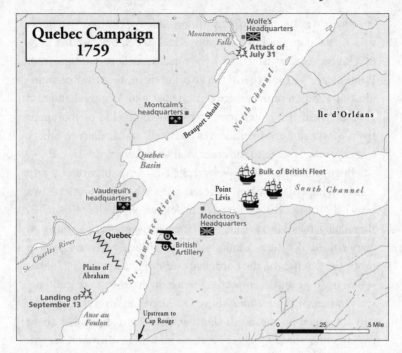

Quebec Campaign 1759

won the day in Gabarus Bay. But the grenadiers were cut down by concentrated French fire.

From then on, the combined attack faltered. Wolfe, himself no stranger to impetuosity, was highly critical of the behavior of the grenadiers; and by the time his troops stumbled back to their ships and the lines east of the Montmorency, he had lost 443 killed and wounded. Governor Vaudreuil was smug, but a little premature, when he intoned, "I have no more anxiety about Quebec."[18]

Anxiety was about all that Wolfe had. Montcalm's lines were stretched so wide and defended so well that Wolfe simply could not get in close enough to the walled city to dig the trenches and place the artillery that were customary in siege warfare. Frustrated and distraught by the dismal affair at Montmorency, Wolfe began to look upriver for alternatives.

As early as the night of July 18, two ships of the line, two heavily armed sloops, and two transports had passed beneath the guns of Quebec under the cover of a furious artillery barrage from Point Lévis. This foray had the effect of disrupting river communications and commerce between Quebec and Montreal and also gave the British a closer look at the northern side of the upper river. Montcalm had fortified this area well, and artillery at Sillery caused some damage to the command ship *Sutherland*.[19]

But operating on the upper river had its merits, particularly with regard to impeding the flow of arms and men between Canada's two principal towns. To that end, on August 4, Wolfe dispatched Brigadier James Murray and 1,200 men upriver aboard a squadron commanded by Rear Admiral Charles Holmes. Murray was to probe the defenses on the north bank west of Quebec while Holmes was to frustrate any attempts the French might make to descend the river in their few remaining ships. On August 8, Murray met some stiff resistance on the north shore at Pointe aux Trembles, about twenty miles above Quebec. Next, he raided Saint Anthony across the river on the south shore and a supply depot at Deschambault farther upstream. French naval forces were content to retreat to the mouth of the Richelieu River draining Lake Champlain.

Perhaps the most welcome news that Wolfe received from Murray and Holmes on their return on August 25 (most of Holmes's squadron remained on guard upriver) was information from several prisoners that General Amherst had prevailed against Ticonderoga and Crown Point. Fort Niagara, too, was in British hands. This was grand news, but why, Wolfe wondered, was there no sign that his commander in chief was striking north to apply pressure against Quebec from the south? The answer would become clear only in hindsight.

Amherst was busy reconstructing Fort Ticonderoga and building his new post at Crown Point. He was also awaiting the construction of a brigantine on Lake Champlain before making his way against the most recent line of French defenses at the Ile-aux-Noix at the lake's outlet. Not having heard from Wolfe since early July, Amherst was wary that Montcalm might yet appear in front of him with a unified army. As for help from Niagara, after learning of Prideaux's death and Sir William Johnson's confused assumption of command there, Amherst dispatched Brigadier Thomas Gage to take command of the newly won western posts. Gage may have been Amherst's most able administrator, but he was never one to move quickly. He was content to secure the British gains and not run the risk of sending troops down the Saint Lawrence against Montreal. Wolfe, then, was on his own, just as he had been from the beginning.[20]

But by now Wolfe was dreadfully ill, his normally weak constitution ravaged by an assortment of fever, consumption, and an inability to urinate except with excruciating pain. The prescribed cures—bloodletting and opiates—did little to improve the clarity of his thinking. Recognizing that "the public service may not suffer by the General's indisposition," he summoned his three brigadiers to a council of war and asked their advice on one of three courses of action. It was the first time that Wolfe had done so, and his own decisions during the past two months had hardly endeared him to Monckton, Murray, or Townshend.

Now, Wolfe proposed three alternatives. First, ascend the Montmorency eight or nine miles and fall on the French rear above Beauport in concert with another attack west from the falls. Second, make a combined amphibious and land attack directly against the Beauport shore. Finally, replay the plan of July 31 and attack from the river and across the ford at Montmorency. Essentially, they were all variations of the same theme, and the brigadiers were not persuaded by any of them.

"General Wolfe's health is but very bad," wrote Townshend to a friend in England. "His generalship in my poor opinion—is not a bit better, this only between us." Even fifteen years later, Murray was still writing to Townshend of Wolfe's "absurd" fixation on attacking the enemy's lines at Beauport.[21]

And even if they gained Beauport, could they cross the Saint Charles against Quebec? The brigadiers were unanimous in their counsel. All favored abandoning the Montmorency camp, transferring operations to the south shore, and moving upriver to force an open field battle with Montcalm by decisively cutting off his lines of communication and supply from Montreal. Reluctantly, Wolfe agreed. Of his failed plans at Beauport, Wolfe confided to Saunders the day after meeting with his brigadiers that "My ill-state of health hinders me from executing my own plan: it is of too desperate a nature to order others to execute."[22]

During the first week of September 1759 more British men, ships, and matériel passed upriver of Quebec. Now, Montcalm had to make a decision. Was this another feint or the real thing? In addition to militia units, the key military force immediately west of Quebec was a 2,000-man contingent of cavalry and foot soldiers under the command of Montcalm's trusted aide-de-camp, Louis Bougainville. This force was charged with shadowing the British fleet as it moved between Quebec and Pointe aux Trembles twenty miles to the west and coming to the aid of the militia units to repulse any attempted landing. For a time, the Guyenne battalion of regulars was also camped outside the city's western gates to assist in these efforts. "I need not say to you, sir," wrote Govenor Vaudreuil to Bougainville on September 5, "that the safety of the colony is in your hands."

Montcalm, too, was concerned with his lines of both supply and retreat west of Quebec, but he was still far from certain that Wolfe would mount a major offensive from that direction. Having waited patiently at Beauport and at the Montmorency most of the summer, Montcalm saw these British movements upriver as just a diversion. Montcalm still believed as late as September 2 that Wolfe, "having played to the left [Montmorency], and then to the right [upriver], would proceed to play to the middle" directly against Beauport. And in fact, Wolfe may well have done this, had it not been for his brigadiers.[23]

By September 7, Bougainville's force was moving from Sillery, just west of Anse au Foulon, and following British movements upriver to Cap Rouge, about eight miles above Quebec. Wolfe issued orders that Monckton would lead a major force ashore at Pointe aux Trembles on the morning of September 8, while other units would make a feint against Cap Rouge to hold Bougainville in place. But the weather turned stormy with torrents of rain and Wolfe countermanded the orders. Instead, some 1,500 troops were permitted off their transports and given a two-day respite on the southern shore.

During this time, Wolfe came up with a scheme much different from that previously discussed with his brigadiers. Taking a boat downstream on September 9 with Monckton, Wolfe spent hours reconnoitering the narrow gap in the cliffs at Anse au Foulon. He made similar observations on the following two days and by September 12 told Monckton that his troops would lead the assault against this point. Monckton was dubious, both because of the French positions atop the cliffs and because of the fact that Murray and Townshend had not been told what their roles might be. Was Anse au Foulon to be the main attack or merely a diversion?

Why and how Wolfe came to focus almost secretively on Anse au Foulon has long been a matter of debate. Wolfe's admirers offer it as a testament to his genius and generalship. His detractors term it a desperate, last-ditch gamble in which Wolfe was determined to die rather than return to England in disgrace. Some say that Wolfe discovered the rocky defile himself, that he was told of it by a French deserter, or even that he learned of it from Captain Robert Stobo, who had achieved fame as a British spy at Fort Duquesne and had recently escaped from Quebec.

The truth may lie in a report from Admiral Holmes, whose ships would bear the burden of the naval operations. Holmes observed a few days after the battle that Wolfe's "alteration of the plan of operations was not, I believe, approved of by many, besides himself." According to the admiral, Anse au Foulon "had been proposed to him a month before, when the first ships passed the town, and when it was entirely defenseless and unguarded, but Montmorency was then his favorite scheme, and he rejected it."[24]

Now, on the evening of September 12, Wolfe's brigadiers sent him a joint letter pleading to be told his specific plans. Wolfe's reply was indifferent, almost condescending: "It is not a usual thing to point out in the public orders the direct spot of our attack." He had given the navy specific instructions for loading troops, into small boats. He himself would be in the van of Monckton's troops and Monckton had already been told that his troops would lead the attack. If they were successful, Townshend was to follow, and presumably, Murray after him. What more did they need to know?

But as Wolfe gave these orders in his cabin aboard the *Sutherland*, it appeared that far from planning a grand victory, he was planning his own funeral. He had summoned a friend, Lieutenant John Jervis of the Royal Navy, who would go on to fame in the Napoleonic wars, and given Jervis his personal papers, a copy of his will, and the miniature portrait of Katherine Lowther. Set the portrait in jewels and return it to her with his compliments, the general commanded. He was dressed in a bright new uniform, and whatever infirmities he had recently suffered were momentarily put aside. Midnight came, and two lanterns appeared in the maintop of the *Sutherland*. It was time to go.[25]

The part of Wolfe's legend that maintains that Anse au Foulon was the only possible spot west of Quebec where a landing might be forced is simply not true. Strategically and tactically, Pointe aux Trembles offered numerous advantages and at the time was guarded by only 190 men. By putting 4,000 troops ashore there, Wolfe could have moved eastward, dispensed with Bougainville's smaller force at Cap Rouge, and completely cut Montcalm's lines to Montreal.

Instead, Wolfe gambled on moving his army down the swift current of the Saint Lawrence in a flotilla of tiny boats, landing at one rocky spot, climbing in single file up a steep cliff, and forming on the plain above before they could be repelled. Perhaps Wolfe was a genius or, as the duke of Newcastle supposedly described him, "simply mad." For his part, Admiral Holmes called the plan "the most hazardous and difficult task I was ever engaged in—the distance of the landing place; the impetuosity of the tide; the darkness of night; and the great chance of exactly hitting the spot intended, without discovery or alarm; made the whole extremely difficult." But, evidently, it was not impossible.

Bougainville was still at Cap Rouge with more than 2,000 men. Admiral Holmes had been toying with him by having his ships float upriver with the high tide and then descend again with the ebb. Weary of following what had become a daily show, Bougainville settled in at Cap Rouge to await more concrete developments. Montcalm was east of Quebec at Beauport, still certain that whatever the British were doing upriver, it was part of an elaborate ruse, and that the main attack would come on the Beauport shore. Admiral Saunders did all that he could on the night of September 12 to confirm this opinion by lowering boats, placing buoys, and unleashing a heavy bombardment against these French positions in an elaborate ruse of his own.[26]

On the evening of September 12, 1759, as Wolfe made his personal preparations, about 4,000 troops embarked in flatboats on the south shore. Other ships made their daily float up to Pointe aux Trembles and then turned with the tide and under a favorable south wind raced back down river to join them. By two o'clock in the morning on September 13, the flotilla of boats, led by Wolfe and Monckton, was moving down the Saint Lawrence toward Anse au Foulon.

French sentries who heard the boats pass and might have spread the alarm were surprisingly gullible. Word had been passed along the river posts to expect a convoy of provisions from Montreal trying to slip past the British fleet in the dark of night. This must be it. When French challenges of "Who goes there?" from the shore were answered "Français," there was no alarm raised.

At four o'clock in the morning thirty advance boats with about 1,800 men touched ashore at the base of Anse au Foulon and several points above and below it. Led by Lieutenant Colonel William Howe—the younger brother of Lord Howe, who had died at Ticonderoga—and followed closely by Wolfe himself, a detachment of light infantry clambered up the slopes to the plain above. If Howe, because of his own personal loss, had any misgivings about the presence of his commanding general in such a dangerous position, he gave no indication of it.

But the surprise was not complete, and when a small outpost of sixty Canadian militiamen opened fire, Wolfe sent orders down to the beachhead not to land any more troops. Was this his death wish or merely confusion again, as when he waved his hat in Gabarus Bay before Louisbourg?

Whatever the answer, it didn't matter. Wolfe's adjutant, Major Isaac Barré, disregarded the order and continued to urge the thin line of troops up the winding path. Meanwhile, Montcalm heard the firing west of the town but sat transfixed on the Beauport shore, thinking that the commotion upriver must be from an attack on his provision convoy. Even an early report that the British were in possession of Anse au Foulon was deemed to be suspect. Just the same, Montcalm ordered the Guyenne battalion, lately posted at Beauport, back to its position just west of Quebec.

Finally, mounting evidence convinced Montcalm that he himself should ride to the sound of the guns and take with him as many troops as he could muster. By the time Montcalm reached the crest overlooking the Plains of Abraham with some 4,500 regulars and militia, Townshend's second wave had come ashore at Anse au Foulon and bolstered Wolfe's force to more than 4,000.[27]

Much has been said by British writers about the glory of Wolfe's "thin, red line" stretching across the verdant Plains of Abraham on that cool, misty September morning. Just as much has been written from the French side to the effect that as Montcalm beheld this, he and all of New France trembled and foresaw their doom. But should they have?

The year before, Montcalm had waited patiently for Abercromby's hideous blunder before Ticonderoga. There was no tangle of abatis to hide behind on the Plains of Abraham, but Wolfe had not even moved his line forward to seize the high ground in the middle of the plain. Having sent a messenger to Bougainville urging him to rush from Cap Rouge and fall on the British rear, why didn't Montcalm wait for him? At the very least, why didn't Montcalm wait to be attacked by British ranks moving uphill? Incredibly, after waiting patiently all summer and refusing to engage Wolfe in a major battle, Montcalm moved with alacrity to do just that. If Wolfe seemed determined to die trying to capture Quebec, Montcalm now seemed resigned to the fact that he must die defending it.

And so, in this most unconventional of wars by European standards, the centerpiece battle was to be fought in the most conventional of European styles—two long lines of opposing forces, one rushing forward, the other content to stand tall and fire away. Once Montcalm had decided to attack, the topography of the plain gave him little option but to do so in a frontal charge.

The British line was stretched from the heights above the Saint Lawrence for almost half a mile to the bluffs above the Saint Charles. At five minutes to ten on the morning of September 13, 1759, about 2,000 French regulars, their ranks thinned by years of service in North America; and some 1,500 Canadian militia started forward with a great cheer. "It was," the historian Fred Anderson wrote, "almost the last thing they would do in unison that day."

At about 150 yards from the British position, the French troops, their ranks already in some disarray, dropped on one knee and fired a volley by platoons into the British line. Standing on a rise on the right flank with the Louisbourg Grenadiers, who had prematurely rushed into action at Montmorency, Wolfe was among the first to be hit, sustaining a shattered wrist. Reloading, the French line moved forward to close the gap. No one in the British line had yet fired a shot.

Onward the French came. At about sixty yards the British right and left flanks opened fire by platoons, but the Forty-third and Forty-seventh regiments in the center gave the onrushing ranks another twenty yards. Then, with the French attackers only forty yards away, the center of the British line exploded in a single volley that staggered the French advance and stopped it cold. As the British reloaded, the French units disintegrated and beat a hasty retreat.

Wolfe, apparently not content to be merely the victor, ordered a bayonet charge in pursuit and fell mortally wounded with wounds in his chest and intestines. Montcalm, too, had fallen, his abdomen and leg ripped open by grapeshot. Wolfe would expire on the field, Montcalm early the following morning.

As the British surged toward the walls of Quebec, their lines and discipline deteriorated, thanks in part to a concentration of fire from the remaining Canadian militia in a cornfield on the French left. Quite suddenly, there was mass confusion. Wolfe was dying or already dead. Monckton lay injured. Barré was incapacitated with a face wound. Murray was bogged down in a fight with arriving reinforcements above the Saint Charles, and the advance units of Bougainville's column were appearing to the south.

Command devolved on Brigadier Townshend, who suddenly found himself in the military role he had sought. Townshend hastily dispatched orders to regroup. Seeing this coalescence begin, Bougainville held up his attack and withdrew to the safety of the Sillery Woods. His arrival an hour earlier, or Montcalm's wait of an hour, might have led to a far different story.[28]

Not surprisingly, Governor Vaudreuil was among the critics who charged that Montcalm had acted "precipitately." But he was not alone. One French officer wrote that "it was the judgment of everyone that had Montcalm awaited the arrival of Bougainville, so that the combined forces could have struck the enemy, not an Englishman would ever have re-embarked." It would have been a classic pincer movement. Instead, it was the classic blunder of largely uncoordinated units of ill-trained militia and regular battalions long removed from the discipline of European battlefields rushing forward against a solid line. Vaudreuil had wanted to use these men to defend New France with frontier tactics across a wide region. Now, Montcalm had assembled them to fight a very different kind of battle and had met with disastrous results.[29]

By one count the British lost sixty men killed and 600 wounded. The French lost 200 killed and 1,200 wounded. The death of each side's commanding general left history with two pithy quotations: Wolfe glad that he could die happy in victory and Montcalm appeased that he would not live to see the English masters of Quebec. The latter, of course, was now about to happen.

Vaudreuil and his militia and Bougainville and his force retired westward toward Montreal. The commandant of the garrison at Quebec, Jean de Ramezay, was left with an assortment of about 2,200 soldiers, sailors, and militiamen and three days of rations with which to defend the town and its 6,000 frightened inhabitants. Vaudreuil had no misconceptions about the outcome and left Ramezay instructions to surrender when his provisions ran out.

But in retrospect, this action by Vaudreuil appears to have been hasty, too. The chevalier de Lévis, who heretofore had been charged with protecting Montreal from Amherst, rallied the retreating troops and marched back toward Quebec with eighty carts of provisions and as many as 5,000 men. This relief force, which may well have played havoc with Townshend's dwindling numbers, was only a dozen miles from Quebec when word reached it that Ramezay had surrendered on the afternoon of September 17.[30]

The following day a detachment of the Royal Artillery and the Louisbourg Grenadiers marched into the Upper Town and raised the Union Jack over the citadel of Quebec. In many respects it was a frighteningly hollow victory. The British immediately granted the French troops safe passage to France with full honors of war—not as prisoners—and permitted the militiamen to lay down their arms and rejoin their families. Most of the regulars were delighted to leave; but far from being magnanimous, the British were being practical. They simply did not have the troops to impose harsh terms. And their numbers were about to grow even smaller.

On October 18, Admiral Saunders and the bulk of the British fleet weighed anchor and sailed for England before another winter's ice claimed the Saint Lawrence. Brigadier Monckton chose to go to New York to recover from his wounds; Brigadier Townshend elected to return to England, no doubt in order to look after his political health. That left James Murray, the junior brigadier, to take on the thankless task of presiding over the winter garrisoning of Quebec.

Perhaps no circumstance emphasized the tenuousness of Wolfe's victory more than the situation in which Murray now found himself. The British fleet was gone, an able French army lurked just to the west, and Quebec was effectively cut off from all aid and succor. The British may have captured Quebec, but what they had really done was to trade their position as attackers and become the besieged. With some justification, a newspaper in Paris reported that "the English hold no more than the ruins of Quebec." This coming winter, it would be British troops who would go hungry and wait to be attacked. Had it really been the battle for a continent? If so, it was not yet over.[31]

And what of the flaming James Wolfe? His body was onboard the *Royal William*, bound for England for burial, but what of his reputation? William Pitt himself set the tone in a speech to Parliament that seems to have surpassed even his usual hyperbole. "Carthage may boast of her Hannibal and Rome may decree triumphs of her Scipio," a newspaper declared, "but true courage never appeared more glorious than in the death of the British Wolfe."

Later, British Canada would sing, "In days of yore from Britain's shore, Wolfe the dauntless hero came, and planted firm Britannia's flag on Canada's fair domain." Wolfe would forever be called "the Conqueror of Canada" But did he deserve the name? Christopher Hibbert subtitled his biography, *Wolfe at Quebec*, "The Man Who Won the French and Indian War." Taken in that light, Montcalm's biography might just as readily be subtitled "The Man Who Lost the French and Indian War."[32]

14

THE MAKING OF A LEGEND

If the fall of Quebec and the death of James Wolfe came to dominate a year filled with banner events, there is one more story that must be told before the curtain falls on 1759. Not surprisingly, the story belongs to Major Robert Rogers and his rangers and it begs to be told, if for no other reason, because of its seminal place in American frontier mythology. Whether in James Fenimore Cooper's nineteenth-century tales or dime novels of the West, a good story has frequently obscured the truth. This occurred all the more in the twentieth century when motion picture images of historical events and personalities became so ingrained in the public psyche that it became difficult, if not impossible, to separate the historic character from the actor. Such is definitely the case with a resolute Spencer Tracy playing the role of Major Robert Roberts in 1940, in the film version of Kenneth Roberts's classic, *Northwest Passage*.

The setting for the original drama was as follows. Amherst's advance down Lake George and his capture of Ticonderoga and Crown Point in midsummer of 1759 had been less a stunning success for British arms than a tactical withdrawal by the French. Neither Fort Carillon nor old Fort Saint Frédéric at Crown Point could have withstood a determined British artillery assault. (Amherst was no Abercromby and doubtless would never have ordered another frontal assault by infantry.) So, under orders from the chevalier de Lévis at Montreal, the French forces on Lake

Champlain retired northward to Ile-aux-Noix near its outlet. Having acted quite decisively until this point, Amherst grew cautious. In some respects, he took the bait and spent the remainder of the summer rebuilding these captured posts rather than advancing toward Montreal.

Besides building forts, however, Amherst was faced with two major uncertainties: the maritime control of Lake Champlain and the status of Wolfe's attack against Quebec. As to the former, the French had a heavily armed schooner and three xebecs (small three-masted ships suited for lake or coastal operations) on the lake. While hardly an armada, this tiny flotilla nonetheless could have caused considerable destruction if turned loose against an advancing convoy of British whaleboats and bateaux. To counter this threat, Amherst had to launch a navy of his own; and as he built forts, his sawmills also constructed a brigantine and a "great radeau or raft, eighty-four feet in length and provided with sails, capable of carrying six twenty-four-pounders." Once afloat, these vessels would provide cover for any convoy, but their construction also meant more delays in any northward movement.

That left the uncertainty of Quebec. If Wolfe was unsure of what role Amherst might play in operations along the Saint Lawrence, Amherst was equally in the dark about Wolfe. Early in August, the commander in chief dispatched a small party consisting of Captain Quinton Kennedy, Lieutenant Archibald Hamilton, and seven Stockbridge Indians to travel through the country of the Abenaki Indians under the cover of a flag of truce. Ostensibly, Kennedy's mission was to make peace entreaties to the Abenaki, but in fact he was trying to obtain some word of Wolfe or even to reach Wolfe's headquarters. This was definitely stretching the use of the white flag, and Kennedy and his party were detained by Abenaki from the village of Saint Francis and turned over to the French as prisoners. When news of their capture finally reached Amherst on September 10, he was furious. Enter Rogers and his rangers.[1]

Rogers had been out and about from Crown Point on various scouting expeditions. About 200 of his rangers under Captain John

Stark had just returned to Crown Point after cutting a road from there through the Green Mountains to the outpost of Number Four on the Connecticut River (present-day Charlestown, New Hampshire). As Rogers remembered it, Amherst, "exasperated at the treatment of Captain Kennedy," summoned him on September 13 and secretly ordered him to attack the Saint Francis Abenaki in retaliation and "bestow upon them, a signal chastisement."[2]

Whatever Amherst's motives in dispatching Rogers on this assignment, "he could not have assigned a duty more agreeable to the rangers." The Abenaki had long been the scourge of the New England frontier. If Rogers's journal is to be believed, some 600 scalps hung from poles at Saint Francis alone. While the rangers were not above similar acts, there was scarcely a man among them who had not had a family member or friend killed or abducted by the Abenaki. The Saint Francis Abenaki professed the Catholic faith, as a result of decades of missionary work by the Jesuits; but far from tempering their actions, this indoctrination made it all the more easy for French priests to suggest that attacks against the English would put them in higher favor with the church. Perhaps their most egregious conduct had occurred in the aftermath of Fort William Henry. To a man, the rangers thought that it was high time for retaliation.[3]

Rogers and some 190 men departed Crown Point in seventeen whaleboats on the evening of September 13, 1759—by coincidence, the day of the great battle at Quebec. Only 132 of this number were true rangers, the remainder being handpicked veterans from the regular light infantry and certain provincial units. Of the rangers, at least twenty-four were Stockbridge Indians and one was a twenty-six-year-old black man, "Duke" Jacobs, who had been granted his freedom after three years with the rangers, but who nonetheless volunteered for the Saint Francis raid.[4]

Dawn the next morning found the force holed up at Button-mould Bay on the eastern shore of Lake Champlain. On their second night out, Rogers and his men reached the mouth of Otter Creek where they were forced to wait until patrolling French vessels moved farther up the lake toward Crown Point. Finally, ten days after leaving Crown Point, the little flotilla arrived at Missisquoi Bay in the northeast corner of Lake Champlain. Here, Rogers hid his boats, and with 149 men—the remainder had already started back to Crown Point with various maladies—set off cross-country for Saint Francis (near modern-day Pierreville, Quebec), about seventy-five miles away as the crow flies.

Two problems quickly presented themselves. First, the men were not flying, but slogging through an interminable spruce bog covered with water a foot deep. Such terrain added immeasurably to the distance, not to mention the discomfort. Second, on the second day in this quagmire, the two Stockbridge Indians who had been left to watch the whaleboats caught up with the main body and reported that the French not only had discovered and burned these boats but also were in pursuit with some 300 men. Barely into the raid, the rangers' presence was known and their planned avenue of escape closed.

Rogers now leaned on his knowledge of the New Hampshire frontier. He sent Lieutenant Andrew McMullen and six others back to Crown Point with a report to Amherst that he intended to make a wide loop eastward and return by way of Lake Memphremagog and the Connecticut River and that Amherst should hurry provisions north from Number Four to meet him. As for their objective, "We now determined to out-march our pursuers," Rogers wrote later, "and destroy Saint Francis, before we were overtaken."[5]

In some respects, the spruce bog became their ally, making any pursuit as tortured as their own advance. Today, when we are used to Gore-tex and weatherproof boots, it is difficult to imagine being wet in wool clothing and leather moccasins for days on end. At night the rangers hacked spruce boughs off trees and stuffed them into the lower branches to make fragile but welcome hammocks above the morass. Food was already a problem.

For ten days they struggled through the swamps and then on the evening of October 2 emerged on the west bank of the Saint

Francis River about twelve miles above the village. As it lay on the opposite shore, the rangers were forced to cross the five-foot-deep Saint Francis the next day by linking arms together in a human chain. By that evening, they were hidden in the woods surrounding the village. Half an hour before sunrise on October 4, 1759, Rogers and 142 of his men rushed forward to attack.

The village of Saint Francis consisted of over "sixty well-built, framed and windowed houses covered with bark, boards and even stone." The two principal buildings were the Jesuit mission church and the council house that had been the scene of a wedding feast the night before. Rogers estimated the population killed in the next hour at 200, but that number is highly suspect. Many of the warriors were not in town, having left previously to await Rogers's emergence from the spruce bog at Yamaska to the west. Others fled and hid with women and children in a little ravine just east of the town. Still others hid in their homes and probably did perish as Rogers's men torched every building in town except three granaries of corn. Rogers took twenty women and children prisoners, and some others were undoubtedly killed in the fires despite Amherst's instructions that "no women or children should be killed." One French estimate gave the number of casualties at Saint Francis as ten men and twenty women and children.[6]

As smoke from the burning buildings filled the autumn morning, many rangers hurriedly scooped up corn and provisions. Others gathered plunder, including, it has long been alleged, numerous icons from the mission. These would become part of a subset of the legend of this raid—the lost "treasure" of Saint Francis. In the grim weeks ahead, most rangers would have gladly traded anything for the corn.

And then, within minutes, it was time to go. Soon, parties of French and Abenaki from Montreal to Three Rivers would be in pursuit. They included sixty regulars and militia under the command of Major Jean-Daniel Dumas, the officer who, as a young lieutenant in 1755, had steadied the French line and halted Gage's advance guard on the Monongahela. Now, there was clearly no

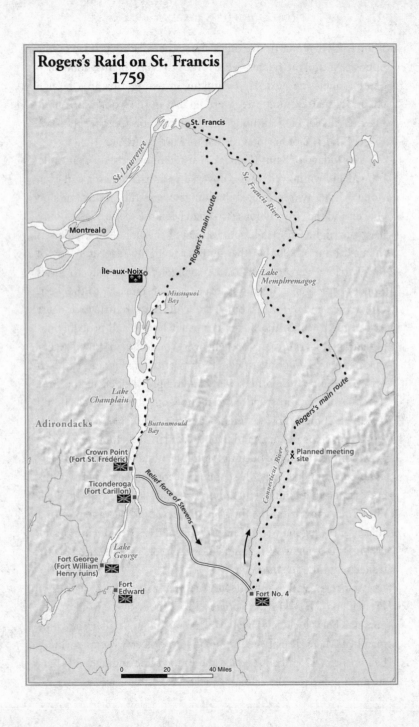

Rogers's Raid on St. Francis 1759

St. Lawrence

St. Francis

St. Francis River

Montreal

Île-aux-Noix

Missisquoi Bay

Rogers's main route

Lake Memphremagog

Lake Champlain

Adirondacks

Buttonmould Bay

Rogers's main route

Crown Point (Fort St. Frédéric)

Ticonderoga (Fort Carillon)

Relief force of Stevens

Connecticut River

Planned meeting site

Lake George

Fort George (Fort William Henry ruins)

Fort Edward

Fort No. 4

0   20   40 Miles

escape for the rangers except up the Saint Francis and down the Connecticut.

For eight days, Rogers led his men up the east bank of the Saint Francis River, averaging about nine miles a day. Then, at the fork of the Magog River (present-day Sherbrooke, Quebec), as the rations of corn from Saint Francis gave out, an officers' council prevailed on Rogers to divide the command into smaller groups in the hope of finding game. Rogers had hoped to keep the party intact for at least another thirty or so miles to the southern end of Lake Memphremagog and the divide between the Saint Lawrence and Connecticut watersheds. But on the morning of October 12, he reluctantly agreed to follow his own ranger rules for dispersal and divide into eleven parties.

Three groups determined to strike southwest directly toward Crown Point, while the others agreed to rendezvous on the Connecticut River about sixty miles above Number Four, near where Rogers had proposed the construction of Fort Wentworth in 1755. Here, they hoped to find the supplies Amherst had dispatched from Number Four.[7]

In his request to Amherst sent via Lieutenant McMullen, Rogers had asked that Lieutenant Samuel Stevens be placed in charge of the relief column because of his knowledge of the upper Connecticut. Stevens had in fact commanded a platoon of rangers at Number Four during the winter of 1758–1759. On the same day as the attack on Saint Francis, Amherst ordered Stevens to gather supplies and men at Number Four, march north to the mouth of the Wells River, and remain there "as long as you shall think there is any probability of Major Rogers returning that way." The question that would haunt Stevens's subsequent court-martial—and the question that is still being asked today among aficionados of Rogers' Rangers—is this. Given Rogers's already legendary record of survival, how long a wait was enough?[8]

Stevens's timing was uncannily close. His relief party of five men from Number Four paddled a large canoe laden with supplies up the Connecticut, portaging several falls, and arrived within five miles of the Wells-Ammonoosuc-Connecticut confluence on October 19. Here, they halted because of strong rapids, but Stevens and others "went daily by land to Wells River and fired their muskets to signal Rogers." But for how many days? Maybe only one.

Rogers reached the appointed rendezvous on October 20. His party numbered twenty-six others. Both Rogers and Stevens reported hearing musket shots. Stevens discovered two hunters, determined them to be the cause, and shortly thereafter paddled back down the Connecticut with the supplies. Rogers discovered Stevens's fire, still smoldering, at Wells River and collapsed in despair.[9]

For Rogers's partisans to this day, Stevens remains the villain, never to be forgiven. Rogers summoned his strength and with three others made an epic descent of the Connecticut by raft to Number Four. Within half an hour of his arrival, supplies were hastened back upriver to the forks, but by one count thirty-four men in the groups that finally reached there died of starvation before help arrived. Other returning parties ended up scattering as far east as Lake Winnipesaukee, and one managed to reach Crown Point directly. Rogers himself rested at Number Four only one day and then returned upriver with two more canoes of provisions.

Rogers was more than bitter. He pushed for Stevens's court-martial the following spring, and the lieutenant was found guilty of neglect of duty and relieved of his command. Stevens had arrived back at Crown Point on October 30, saying that it was most unlikely "that Rogers would return by way of Number Four." Amherst himself expressed some skepticism and recorded in his journal that Stevens should have waited longer. Eight days later, Captain Amos Ogden, who had rafted down the Connecticut with Rogers, walked wearily into Crown Point with Rogers's report to prove the contention.

If Stevens has any supporters, they remain cowed by the insurmountable legend of Rogers and his rangers. It might only be said in Stevens's defense that waiting with only five others—townsmen no doubt eager to return to Number Four—along the route of known Abenaki incursions into New Hampshire must have given Stevens pause. But why didn't he at least cache provisions at his northernmost point? The answer is that he had explicit orders not to do so. Amherst's direct orders were that after waiting at his own discretion, Stevens was to "return with said provisions and party to Number Four." Likewise, Amherst's orders to the commissary at Number Four indicated that if Stevens did not encounter Rogers, he would "return with said provisions to you, and you will return them into the King's stores." Even if Stevens had considered doing otherwise, he was no doubt swayed by his belief—as he testified at his court-martial—that if he was going to meet Rogers at all, it would have been earlier at Number Four.[10]

By the best estimates, Rogers lost 69 of the 142 men who left Missisquoi Bay with him. One was killed in the attack on Saint Francis; seventeen died in various ambushes on the return; forty-three succumbed to starvation; and the fate of eight is unknown. Even compared with the most horrendous encounters of the period, this casualty rate was appalling. There is some evidence that both Rogers and General Amherst sought to sweep the figures under the table or at the very least not tally them all in one place.[11]

But it hardly mattered. In many respects, the numerous deaths and the hardships of the retreat only heightened this exploit in the public mind. Contemporary colonial newspapers continued to sing the praises of "the famous Rogers." Rogers himself published his journal and various other accounts after the war, partly to obscure less heroic ventures, as shall be seen some pages hence.

But no one remembered that. Instead, tales of the rangers' campsites, buried treasure, elusive escape routes, and escapades real or imagined permeate upstate Vermont and New Hampshire. Indeed, local stories that "Rogers' Rangers passed through here" became almost as ubiquitous as signs saying "George Washington slept here." By the time Spencer Tracy marched his way across the silver screen in 1940, the legend of Robert Rogers and his rangers was forever intact.

15

DECIDING THE FATE

While London and all of the British Empire gloried in the capture of Quebec, the garrison condemned to spend the winter of 1759–1760 there found little to celebrate. "We are masters of the capital, it is true," wrote one British junior officer, "but it does not follow from thence that we have conquered the whole country, that entirely depends on our fleet." And now the fleet was gone. Brigadier James Murray's 7,000 troops were alone in the middle of a foreign country, outnumbered by still competent enemy forces, and surrounded by hostile elements. "A severe winter now commenced while we were totally unprepared for such a climate," Ensign James Miller recorded, leaving "neither fuel, forage, or indeed anything else to make life tolerable."[1]

Indeed, the only constants became a lack of food and daily burials from a host of diseases. Dysentery and scurvy, in particular, were rampant. Troops were sent out in small parties to forage the surrounding countryside for food and firewood, but in the face of heavy snows, bitter cold, and frequent attacks by French troops it was grim work. By spring Murray counted 1,000 of his garrison dead and 2,000 "totally unfit for any service." Bougainville could write without exaggeration that "the English hold only the outer walls and the King still holds the colony."[2]

But the winter was hard on the French and Canadians as well. Some inhabitants of Quebec stayed there and endured the same deprivations as their British landlords. Most fled to surrounding towns or upriver to Montreal. This flow of refugees taxed Montreal's resources, but it also provided the chevalier de Lévis with a source of manpower to bolster his forces. What he really sought, however, was the same thing that Canada had always needed—help from France.

In late November, several French ships that had taken refuge above Quebec the previous summer slipped down the river just before it froze and ran past the British guns. They carried plaintive missives to the French court from both Lévis and Governor Vaudreuil requesting reinforcements and supplies as early in the spring as possible. Such aid sailing from France by the end of February and arriving "in advance of the English," Lévis asserted, could still "make us masters of the river."[3]

All winter there were rumors in the British garrison that Lévis was about to sweep down the Saint Lawrence from Montreal and attack Quebec in force. By spring, not knowing what the result of his pleas to France would be, Lévis resolved to do just that. On April 8, 1760, the ice went out in the Saint Lawrence below Quebec. By the middle of the month the water was clear upstream to Montreal. On April 20, Lévis started downriver from Montreal with two frigates and assorted ships and bateaux carrying a force of some 6,900 troops, including eight regular battalions; twenty marine companies; 3,000 militia; and 400 Indians. Four days later, by the time he had reached Pointe aux Trembles and landed there—as Wolfe might have done—local militiamen swelled his ranks to 8,000.

Murray was hardly caught unawares, but in the face of such numbers his initial line of outposts at Cap Rouge and straddling the Sainte-Foy road had little option but to fall back. By the evening of April 27, Lévis had advanced to a line just west of the battlefield of September 13, stretching from Sainte-Foy on the northwestern edge of the Plains of Abraham south to the Sillery Woods.[4]

With one eye on the Saint Lawrence searching for any sign of British relief ships and the other on Lévis's movements, Murray now faced a quandary. If he allowed Lévis to move forward and seize the crest of the plains where Montcalm had formed his battle line the autumn before, there was no doubt that Quebec would be besieged at close range. Could the garrison hold out until help arrived? Alternatively, Murray could move out from behind Quebec's walls and establish a defensive line studded with artillery on that same high ground of the Plains of Abraham.

Having apparently learned nothing from Montcalm, Murray chose the latter course and on the morning of April 28, 1760, marched out from behind the walls of Quebec to meet the French threat. "The enemy was greatly superior in number, it is true," Murray later confessed in a report to Pitt, "but when I considered that our little army was in the habit of beating that enemy . . . I resolved to give them battle . . . [and] we marched with all the force I could muster, namely, three thousand men, and formed the army on the heights."[5]

On the plains now it was a scene far different from the sweep of emerald-green of the previous fall. In most places mushy spring snow a foot deep still covered the ground. The size of the French force was not immediately apparent, as some of the men were sheltered in the Sillery Woods on the left of the British and others were still in camp at Sainte-Foy. Having established his line on the crest and spread about twenty artillery pieces along it, Murray should simply have waited, but he showed no more patience than Montcalm had shown the year before. Noting French units on his right moving forward and still in the process of deploying from a column into a line formation, Murray chose to open the battle by an attack against these troops.

The initial British attack met with some success, as did a similar thrust against the French right; but the deep snow and seemingly bottomless mud slowed the British advance and gave the French time to react. Once the bulk of the French troops emerged from the Sillery Woods and got into action, their line stiffened and began to advance on the British left flank, threatening to turn it and come between the bulk of Murray's force and the walls of the city. Because Murray had moved forward rather than waiting along his line of artillery, his field guns were now of little use—the entire battle line had shifted. Attempts to move the guns forward through the snowy mush in the heat of battle proved a disaster. "Our cannon were of no service to us," a British officer lamented, "as we could not draw them through the soft ground and gulleys of snow three feet deep."[6]

Murray quickly realized that his command was in danger of being surrounded, and under cover of disciplined volleys from a rear guard was barely able to retreat into the city. As had been true the previous autumn, a spirited advance by the attackers might have carried Quebec then and there. But faced with this rearguard fire, the French line halted, content to take possession of Murray's artillery, which had been hastily spiked and left mired on the field.

This second battle of Quebec, also called the battle of Sainte-Foy, was a far costlier encounter than the storied affair of September 13, 1759. Murray lost almost one-third of his command: 259 men killed and 829 wounded. The French losses amounted to 193 killed and 640 wounded. Murray had not only "sustained heavier casualties, abandoned his artillery, and retreated," but also lost a much higher proportion of his effective men—28 percent as opposed to less than 12 percent. Thus, Fred Anderson has noted, "it is no exaggeration to say that he had taken a spectacular gamble and sustained a spectacular loss."[7]

On the following day, Lévis began to dig siege trenches and erect artillery batteries within 600 yards of the city. If he could recapture Quebec quickly, all of Wolfe's triumphs of the year before—real or imagined—would have been for naught. Elsewhere, Amherst was still stalled on Lake Champlain and Gage had not dared to descend the Saint Lawrence. With Quebec back in French hands, the British would have to start all over again on the Saint Lawrence. French Canada might have bought itself another year. Knowing these things only too well, Brigadier James Murray—who suddenly looked rather shabby in comparison with James Wolfe—took the only option still open to him. He frantically scanned the river for any sign of a ship and hoped desperately that its flag would be British.

Even before news of Wolfe's capture of Quebec had reached Paris, the French war minister, Choiseul, had been hard-pressed to come up with a way to reverse France's fortunes. What was needed was a plan to save the French colonies worldwide; support France's ally Austria, against Prussia; and otherwise relieve the pressure on French armies on the continent. Choiseul thought about it and realized that he had but one choice—a grand invasion of England. This was not as ridiculous an idea as it might at first seem. Great Britain's regular army was spread throughout the world. If France could mount a cross-Channel invasion, its regulars might march into London unopposed save for a few home guard militiamen. To be sure, France still had the army for such a venture, but did it have the navy?

By the summer of 1759, thanks to Great Britain's furious ship-building, the Royal Navy could float 113 ships of the line in fighting condition. France's navy could muster barely half that number, but what if these ships could be concentrated off France in such a way as to secure, however temporarily, control of the English Channel? Britain's fleet, after all, was sailing in almost every part of the globe.

Choiseul first proposed to ship an army of 20,000 troops out of ports in Brittany onboard ninety transports convoyed by six ships of the line. This armada would make for the Atlantic as if sailing for North America, but then circle northward around Ireland and suddenly appear in the Firth of Clyde to attack Glasgow and Edinburgh. (Some people in Catholic France still held out the hope that Scotland would rally to the memory of the Stuarts and rise up in rebellion.) With Pitt's government suddenly focused on Scotland, another French army would cross the Channel directly to England in a convoy of flatboats guarded by a concentration of at least thirty-five to forty ships of the line. These troops would march on London and seize the economic hub of Pitt's war effort before he knew what was happening.[8]

It was actually the daring sort of plan that Pitt himself might have championed, but Choiseul's first problem was to join the French Mediterranean fleet from Toulon with the Atlantic fleet at Brest. The command of Great Britain's Mediterranean fleet rested with Admiral Edward Boscawen, as his reward for his role in the capture of Louisbourg the year before. For much of the summer Boscawen was successful in blockading Admiral de la Clue's French squadron in the harbor at Toulon. Finally, when Boscawen was forced to retire to Gibraltar for provisions and repairs, de la Clue sailed from Toulon on August 5, 1759, with twelve ships of the line and three frigates and made for the Strait of Gibraltar.

On the evening of August 16, 1759, de la Clue's ships were making their way past Ceuta at the eastern entrance to the strait when the British frigate *Gibraltar* discovered them. The frigate raced for Gibraltar harbor, where Boscawen's ships and crews were in various states of readiness and the admiral himself was dining leisurely with the governor of nearby San Roque.

Meanwhile, knowing that he had been spotted, de la Clue ordered all lights extinguished and with a strong east wind raced westward through the strait and then turned north toward Cape Saint Vincent. But sunrise brought a shock. Instead of a fleet of fifteen, there were but six ships of the line following his flagship. It is easy to say in hindsight that de la Clue should have put on all sail and hurried toward Brest then and there; but when lookouts reported masts to windward, he instead lowered his sails and paused to wait for what he assumed was the rest of his squadron. By the time the number of sails bearing down on him exceeded eight, de la Clue knew that he had been caught.

With a Herculean effort, Boscawen had put to sea in barely two hours with eight ships of the line and was soon followed by six more. Despite a valiant rearguard action by the French seventy-four-gun *Centaur*, the larger British fleet closed with the seven French ships and engaged in a running fight that lasted for the better part of two days. When it was over, five of de la Clue's ships were captured or sunk. Only two escaped and eventually reached

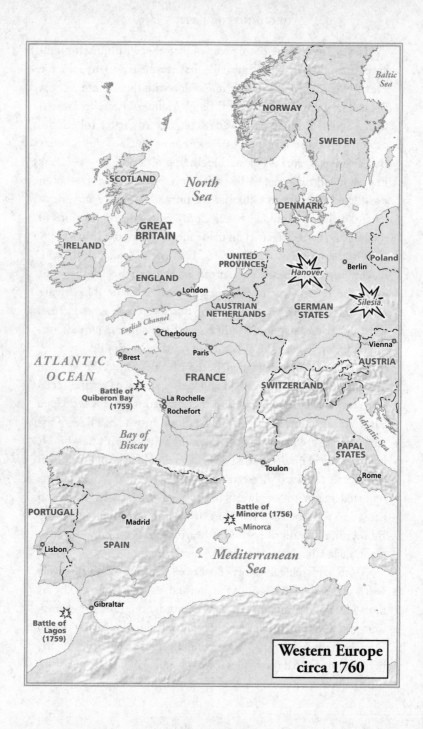

Western Europe
circa 1760

Rochefort in France. And what of the five other ships of the line that de la Clue had waited for at his peril? Along with the three frigates, they had sought refuge in the neutral port of Cádiz, Spain. Recriminations would fly for years over whether their captains had done so under orders or out of cowardice, but whichever, the result for Choiseul's invasion plans was the same. There would be no help from the Mediterranean fleet.[9]

Meanwhile, the French Atlantic fleet had hardly been having its way along the coast of France. Pitt's naval resources had swelled so much that he could send fleets to Canada and the West Indies and still maintain adequate forces in the English Channel. Admiral Herbert de Brienne, comte de Conflans, and the bulk of the French Atlantic fleet were blockaded at Brest. Their nemesis was Admiral Sir Edward Hawke.

Given his proximity to England, Hawke had maintained a continuous blockade by simply rotating a few ships at a time back to English ports for fresh supplies. When six French ships of the line were ordered to break loose from Brest and sail south 100 miles to Quiberon Bay to join up with the transports intended for the diversion against Scotland, they were beaten back into port. Now it became quite clear to Conflans—and was in fact so ordered by Louis XV—that if the planned invasions were to go forward, Conflans would have to sortie his entire fleet, break Hawke's blockade, and rendezvous in Quiberon Bay.

Nature momentarily intervened on the side of the French. Strong autumn gales blew Hawke's fleet off station and forced it to take shelter at Torbay on the southern coast of England. These same gales, however, blew Admiral Bompar's French squadron, which was returning from its participation in the Martinique-Guadeloupe campaigns, safely into Brest. Augmented by Bompar's ships and momentarily without his blockading shadow, Admiral Conflans sailed from Brest on November 14, 1759, with twenty-one ships of the line and five frigates. It was arguably the most powerful French fleet yet assembled in the war. Conflans intended to destroy the weaker British squadron under Commodore Robert Duff that was blockading Quiberon and then join the French invasion transports. Two days later, Hawke's main fleet started south from Torbay under full sail.

The whims of nature changed against the French, and light winds forced Conflans to make a wide circle westward and then spend a day becalmed. Not until the morning of November 20 did he sight the sails of Duff's squadron off Quiberon and give chase. But as the French fleet closed with Duff's ships, its lookouts sighted a far greater cluster of sails bearing down on them from another quarter. Conflans couldn't believe it. It was Hawke with twenty-three ships of the line.

In truth, the forces were fairly evenly matched in firepower, but the boldness of the British navy was about to show itself. Had Conflans been Drake or even Hawke, he might have formed his lines and engaged in one of history's greatest naval battles. Instead, he hoisted as much canvas as his masts could hold and ran for the safety of the French shore batteries in Quiberon Bay.[10]

"I had no ground for thinking," Admiral Conflans wrote later, "that if I got in first with twenty-one of the line the enemy would dare to follow me." But that is exactly what Hawke and his lieutenants did. Fighting fierce squalls, surging seas, and hidden reefs that were as much the enemy as were the French men-of-war, the British first pounced on the rear of the French fleet and cut it to shreds, battering the *Formidable*, *Héros*, *Juste*, *Thésée*, and *Superbe*.

Conflans tried to form a battle line and return to their aid, but the narrow confines of the bay disrupted his efforts. For his part, Hawke eschewed a battle line and boldly hoisted the signal "general chase," in effect turning his captains loose to engage in individual actions. Few ships did so more aggressively than the ninety-gun *Magnanime* under the command of Richard Howe, whose older brother had fallen at Ticonderoga and whose younger brother had just led Wolfe's way at Quebec. "Had we but two hours more daylight," Admiral Hawke reported to the admiralty, "the whole [French fleet] would have been destroyed or taken."[11]

By the following morning, Admiral Conflans was embarrassed

to find that in addition to the losses suffered by his fleet, in the darkness and storms he had anchored his flagship amid Hawke's ships. Trying to slip away, the *Soleil Royal* ran aground and Conflans ordered it burned. In all, the French lost six ships of the line and suffered about 2,500 casualties. British losses were two ships and less than 400 men. More importantly, a dozen of the remaining French ships sought refuge up the shallow Vilaine River and would never emerge as a fighting force. It no doubt seemed an understatement when one French observer wrote: "Imbecility, ineptitude, blundering, ignorance of maneuvering and of all sea tactics are the exclusive causes of our loss."[12]

"The battle of November 20, 1759," wrote Admiral Alfred Thayer Mahan more than a century later, "was the Trafalgar of this war." The French navy, heretofore frequently outgunned but never completely overwhelmed, finally ceased to be a major strategic force. "The English fleets," Mahan continued, "were now free to act against the colonies of France . . . on a grander scale than ever before." Pitt had said that the war would be decided in North America, but in retrospect it was decided in the Atlantic Ocean off the coast of Europe by Boscawen's pursuit off Gibraltar and Hawke's daring in Quiberon Bay.[13]

The French court received Lévis and Vaudreuil's pleas to aid Canada and professed a full measure of support. But by the time this intent was acted upon, a convoy of five merchantmen loaded with only 400 troops and guarded by one lone twenty-eight-gun frigate was all that managed to sail from Bordeaux on April 10, 1760. Arriving in the Gulf of Saint Lawrence in mid-May, the commander of this tiny flotilla captured a British ship and from letters onboard learned that he was following a British squadron into the river. Indeed, a few days before, on May 9, 1760, the masts of a lone ship had appeared in the Saint Lawrence River below Quebec. A close inspection by telescope told the story. It was the frigate *Lowestoft*, and it was the vanguard of a long-awaited relief force. The flag flying from its mast was the Union Jack.[14]

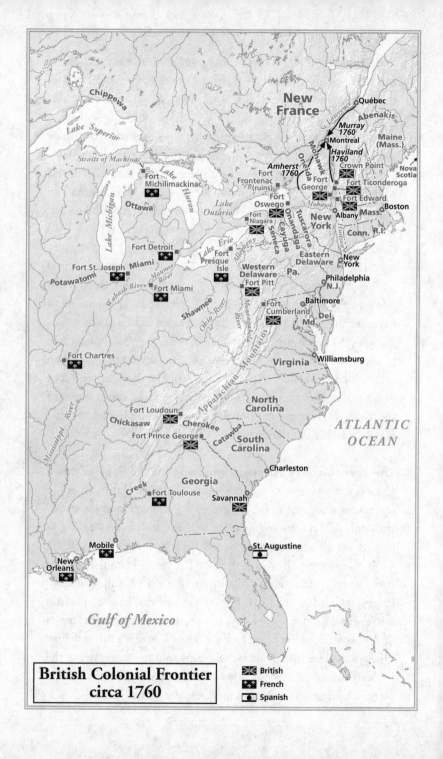

British Colonial Frontier
circa 1760

British
French
Spanish

BOOK THREE

Prelude to Revolution

(1760–1763)

Nothing can eradicate from the English colonists' hearts
their natural, almost mechanical affection to Great Britain.

— GOVERNOR THOMAS POWNALL OF MASSACHUSETTS, *1763*

16

MONTREAL TO
MICHILIMACKINAC

As far back as Braddock's roads in the spring of 1755, the British had always planned multipronged attacks against Canada. Pitt's ascendancy resulted in detailed instructions for the multiple fronts of 1758 and 1759. Now, in the spring of 1760, however, Pitt's orders to Amherst were decidedly succinct. By whatever routes and whatever means he chose, the commander in chief was to advance on Montreal and seize the last major objective in New France.

To be sure, there were some in Canada who still held out hope for a miracle despite the preponderance of British power arrayed against them. "Ah, one ship of the line," lamented one of the French attackers who had briefly surrounded Murray's beleaguered garrison, "and the fortress was ours."[1] That is an interesting "what if," but at best, it seems that if the French had indeed recaptured Quebec, this would have merely prolonged their agony. Amherst's own "what if" is that had he decisively pushed northward in the late summer of 1759, rather than stopping to rebuild Ticonderoga and Crown Point, Montreal might have fallen then and there, while the bulk of French forces were occupied around Quebec.

Now, beginning anew against Montreal, Amherst adopted his

own three-prong plan. Brigadier James Murray was to move south-west up the Saint Lawrence River from Quebec. Brigadier William Haviland was to force Ile-aux-Noix at the outlet of Lake Champlain and strike north down the Richelieu River. Amherst himself would follow Bradstreet's route up the Mohawk to Lake Ontario and then descend the Saint Lawrence northeast to Montreal. Chevalier de Lévis and his remaining forces—a few thousand French regulars and a fluctuating number of militiamen—would be caught in a vise or swept aside.

The sheer numerical superiority of the British on these multiple fronts was staggering: 5,600 regulars and 5,400 provincials with Amherst; 1,500 regulars and 1,900 provincials with Haviland; and 3,800 regulars with Murray, counting 1,000 reinforcements received from Louisbourg. These numbers, threatening from multiple directions, "made it impossible for the Franco–Canadian staff to adopt any fixed plan."[2]

Of these three avenues, Amherst considered Murray's the "easiest." Murray agreed. Indeed, the arrival of the British fleet seems to have given the battered brigadier and his troops not only a new lease on life, but also a conquering edge. Noting that he had not heard from Amherst and fearing that Amherst's movements northward were delayed, Murray wrote on July 14, 1760, that "the Saint Lawrence was the sure route" to Montreal. "I shall push without hesitation to that capital," Murray continued. "I can do it with safety, because I am master of the river, and if Amherst does not get through, which I much doubt, I shall conquer the country."[3]

Whatever his pretensions, Murray got started first. This was in large part because he indeed had the advantage of river trans-port on the Saint Lawrence and did not have to wait for the arrival of provincial troops—an annual ritual that frustrated the advances of Amherst and Haviland until late July. Relying on British sea power, Murray moved slowly up the Saint Lawrence. He issued proclamations calling on the inhabitants to resist at their peril, dispatched raiding parties to show that he meant it, and—despite his earlier boasts—shrewdly refused to land and engage Lévis's remaining troops in force until he was confident that Amherst and Haviland were close at hand.

Meanwhile, Haviland advanced on Lake Champlain and forced Bougainville to evacuate the French works at Ile-aux-Noix. Amherst made similar progress against the defenses at Fort Lévis on the Saint Lawrence and then ran the rapids downstream to the western gates of Montreal, arriving there on September 5. By then, Murray was ashore in force just east of the city.[4]

Governor Vaudreuil and the chevalier de Lévis were out of options. There had never been any good ones—save perhaps prevailing against Murray at Quebec. Taking on the British forces one at a time as they converged on Montreal was not impossible, but it was highly impractical without naval control of the river. There was some talk of a grand retreat westward to Detroit and then south down the Mississippi to the Illinois country and to Louisiana, but that, too, did not come to pass.

With the Canadian militia melting away and the few Indian allies that remained not eager to fight those Indians who now supported the British, the French regulars were left to their own devices. On the evening of September 6, 1760, a meeting of civilian and military leaders was held at Vaudreuil's house in Montreal. All agreed that "a capitulation honorable to the troops and advantageous to the people of Canada was far preferable to an attempt at defense that could only defer the loss of the city for a matter of a day or two."[5]

Consequently, early the next morning, the trusted Bougainville was sent to Amherst's headquarters under a flag of truce with a letter from Governor Vaudreuil. Surrounded though he was, the governor tried to buy a little more time by suggesting a truce until the return of a courier who had recently been dispatched downriver to ascertain whether or not France and Great Britain had come to peace in Europe. Perhaps Amherst suppressed a smile, but if so, he was not amused for long. "Tell Monsieur Vaudreuil," Amherst replied icily to Bougainville, "I have come to take Canada and I will take nothing less."[6]

If the governor had any proposals to make, Amherst continued, he should reduce them to writing. Bougainville mumbled a reply that such documents took time, but when he returned to Amherst's camp by noon the same day with Vaudreuil's proposed terms, it was clear that the governor had taken considerable time over the last few months to craft them in anticipation of the negotiation. Now, Vaudreuil proposed no less than fifty articles to govern the capitulation of not only Montreal but also all of Canada.

Surprisingly, Amherst agreed to many of them—mostly those detailing the administration of the colony and the property rights of its citizens. Some provisions, however, Amherst flatly refused. First, the French regulars would not be accorded the honors of war, but must lay down their arms and return to France with the promise not to serve again in the current conflict. Second, all Frenchmen and Canadians who decided to remain in Canada were free to do so, but they would be regarded not as neutrals—as Vaudreuil had proposed—but as subjects of Great Britain. Third, no sanctuary would be provided to those who had deserted the British army and fled to Canada; these deserters would be dealt with expeditiously. Fourth, in the event that Canada should become a British possession—and there was great uncertainty about that at this point—Vaudreuil had sought assurances that the French king would still be permitted to appoint the bishop of Quebec. This too the general of the Church of England, of course, declined.

Given the circumstances, one of Bougainville's lieutenants thought that "Amherst accorded conditions infinitely more favorable than could have been expected in our circumstances." But for a moment it appeared that Lévis might fight on after all. Most egregious to him was his soldiers' status as prisoners of war and the prohibition against their future military service not just in North America, but in Europe as well. Limit the prohibition to Canada but not Europe, Lévis appealed. He did so because unless his officers were free to bear arms again in Europe, they would be reduced to half pay and would most likely meet financial ruin. Doubtless Amherst knew this, but he remained firm. He did so because of "the infamous part the troops of France had acted in exciting the savages to perpetuate the most horrid and unheard of barbarities." Fort William Henry had finally been revenged.

Lévis threatened to retreat to an island in the Saint Lawrence and fight on, but Vaudreuil, relieved at the terms granted Canada and never too concerned about the fate of French regular officers, convinced him otherwise. By sunset on September 8, 1760, Amherst and Vaudreuil exchanged signed copies of the terms of surrender. Amherst immediately ordered Major Isaac Barré, who had served as James Wolfe's adjutant on the Plains of Abraham and lost an eye there, to depart immediately and carry the news to Pitt. It was Amherst's fondest hope that he, too, would soon be bound for England.

The following morning, the remnants of ten French battalions stacked their arms and returned two British flags taken from Oswego four years before. But where were the French battle flags? Amherst's terms required the surrender of these colors—symbolic of their bearers' status as prisoners of war—but now it appeared that Lévis had managed one small victory. Colors, the French asked incredulously? Why, they were so torn to pieces after six years that they had been destroyed. Undoubtedly, the destruction had been very recent and calculated to prevent them from falling into British hands. Amherst, not one to question the honor of an opposing officer, let the matter drop. The French could keep their colors; he would keep Canada.[7]

The French court took the news of the surrender of Canada with the same uninformed air and detachment that it had exhibited toward Canada's affairs on far too many occasions during the past six years. Professing astonishment that Montreal had opened its gates "without having fired a shot," one official in the ministry of the marine claimed that Amherst's troops should have been no match at all because they had "come over the Niagara Falls." And what of Louis XV? Did he finally have any misgivings that his mistress had lowered his eyes to Europe instead of opening them to the world? Legend has it that the French king was talking with Voltaire when he received the news. "After all, sire," spoke the philosopher sooth-ingly, "what have we lost—a few acres of snow?"[8]

Meanwhile, although Canada might be secure to British arms, there was trouble in the south. There is no doubt that Great Britain and France had long been transfixed by the campaigns along the Saint Lawrence and the Great Lakes. (So, too, would be many historians of the conflict.) It will be remembered, however, that the French encirclement of the British colonies extended all of the way down the Mississippi to New Orleans, east to Mobile, and then northeast to Fort Toulouse at the confluence of the Alabama and Tallapoosa rivers (just north of present-day Mont-gomery, Alabama). Good relations with the Indian nations inhab-iting the buffer zones not only between the British Carolinas and French Louisiana but also between British Georgia and Spanish Florida were essential to the relative stability of these frontiers.

It will also be remembered that political rivalry between Vir-ginia's lieutenant governor Robert Dinwiddie and South Car-olina's governor James Glen had foiled Braddock's attempts to recruit Cherokee warriors from this region to aid in his march against Fort Duquesne. When Attakullakulla (Little Carpenter) of the Cherokee had in fact led warriors north to aid General Forbes in his similar mission in 1758, they had been left with little to do and returned to the Carolinas largely disgruntled.

In November of that year, Governor William Henry Lyttelton of South Carolina wrote to Pitt and offered to mount an attack against French settlements on the Alabama and Mississippi rivers, beginning with Fort Toulouse. In some respects, Fort Toulouse was to the Carolina frontier what Fort Duquesne had been to the frontier of Pennsylvania and Virginia. Such an attack would be popular with his colony, the governor maintained, and he could raise a force of several thousand to accomplish it. Not only would the expedition evict the French from this territory, but it also would serve to impress the Creek Indians there and if not gain their support, at least keep them neutral. In this effort, Lyttelton assumed that he would have the support of friendly Catawba and Chickasaw as well as Attakullakulla's Cherokee. Having boasted of South Carolina's willingness to undertake this mission, the governor concluded his letter to Pitt with numerous requests, including tents for 2,000 men, 2,000 muskets and ammunition, and a direct mail service from Charleston to London.[9]

Even as Lyttelton's proposal made its way to Pitt, Amherst himself had pondered a campaign against Fort Toulouse in the winter of 1758–1759. Nothing came of this, however, because Amherst worried that such troops would be needed in the north for the spring campaigns before an attack against Toulouse might be concluded. Amherst confessed to Lord Ligonier that such an effort would be "lopping off a branch, when it would be time that I should try at the stem and the root." Pitt, for his part, of course, was also focused on Canada and not inclined to open a southern theater of operations. Lyttelton's request was not approved, and South Carolina "fell back into its previous attitude of indifference to support of the war."

Its neighbors were equally indifferent. North Carolina had sent three companies to aid Forbes in 1758, but declined to provide any support for the following year's campaigns. Tiny Georgia had only a small troop of rangers maintained at the governor's expense. All this was to change with what the colonials came to call the Cherokee War.[10]

The Cherokee numbered about 10,000. They lived in forty towns clustered in three main areas along the Carolina-Georgia frontier: the Lower Towns east of the Great Smokies, the Middle Towns tucked among their hollows, and the Overhill Towns west of the mountains. Despite Attakullakulla's generally pro-British stance, there was a strong pro-French faction among the Cherokee that was hardened by the whispers of French traders from Fort Toulouse and by increased resentment over four principal British settlements deep in Cherokee country.

These settlements were Long Canes Creek and Ninety-Six in western South Carolina; Fort Prince George on the Keowee River in extreme western South Carolina; and Fort Loudoun on the Little Tennessee River on the western side of the Smokies, about thirty-five miles southwest of present-day Knoxville. Fort Loudun had been established in 1756 near the Cherokee capital of Chota and was the "first English fort on the western side of the Appalachians." It hardly helped matters that a number of Cherokee women, "enjoying great personal freedom, soon were living with many men of the garrison."[11]

Friction intensified when some of Attakullakulla's warriors returning from the Forbes campaign were mistaken for the enemy and were killed indiscriminately by militiamen. To add insult to injury, other returning Cherokee found that in their absence, colonials had encroached on the hunting grounds of the Lower Towns and poached game vital to the winter hunt. Attakullakulla became the mediator in a rapidly deteriorating situation and tried to salvage relations by emphasizing the beneficial trading relationships that were crucial to both the Cherokee and the South Carolinians.

But when certain Cherokee went to Charleston to further negotiations, Governor Lyttelton not only cut off all trade goods—including the time-honored presents—but also took some of the group as hostages, demanding in return that any Cherokee who had ever killed a settler be turned over for punishment. When this didn't produce results, Lyttelton marched the South Carolina militia to Fort Prince George along with the innocent hostages.

Such saber rattling, of course, only made matters worse. The Cherokee refused to surrender the accused parties, and Lyttelton continued to hold the innocent hostages and some three tons of presents—woolens as well as powder, balls, and muskets. Then, an outbreak of smallpox caused panic among Lyttelton's militia and the governor beat a hasty retreat to Charleston. He left behind a recipe for disaster: a smallpox epidemic, vengeful Cherokees who would never surrender to British justice, resentful relatives of the innocent hostages, and a stockpile of guns and ammunition. Lyttelton's ineptness had lit the fuse, but within two months he departed Charleston to become governor of Jamaica and left the ensuing mess to Lieutenant Governor William Bull.

Isolated, undermanned, and holding both prisoners and booty, Fort Prince George became a ripe target. On February 16, 1760, the Cherokee leader Oconostota tried to lure the fort's commandant into an ambush under the guise of a parley. The commandant was killed, but Fort Prince George held and the innocent Cherokee hostages were murdered in retaliation. This act unleashed Cherokee attacks all along the Carolina frontier, effectively pushing it eastward almost 100 miles.

Now, rather than offering to lead an attack against the French, South Carolina appealed for British regulars to secure its own interior. Amherst responded by dispatching the First and Seventy-seventh regiments of Highlanders under the command of Colonel Archibald Montgomery to Charleston. Montgomery's second in command was Major James Grant, recently exchanged as a prisoner of war after having been captured outside Fort Duquesne. Whether Grant had learned anything about Indian warfare remained to be seen.

Montgomery marched to the relief of Ninety-Six and then moved westward to Fort Prince George, destroying a succession of Cherokee villages in the Lower Towns in the process. "The neatness of those towns and their knowledge of agriculture would surprise you," Grant wrote to Lieutenant Governor Bull. "They abounded in every comfort of life, and may curse the day we came upon them."[12]

From Fort Prince George, Montomery attempted to advance on the Middle Towns where many Cherokee from the Lower Towns had taken refuge. But this was tougher country. The paths were impassable to large wagons, and even the Scots, who were used to mountainous terrain, found the going difficult. On June 27, 1760, near Echoe, a Cherokee attack inflicted 100 casualties and the loss of so many pack animals that Montgomery decided that he had had enough. Abandoning any thought of pushing on to Fort Loudoun, he retreated to Fort Prince George and in short order to Charleston itself.

By August, Montgomery's force was en route back to New York, and Amherst was calling the venture "the greatest stroke the Indians have felt." It was hardly that. Despite leaving some provisions and reinforcements at Fort Prince George, Montgomery had returned to Charleston without addressing any of the underlying issues. For their part, the Cherokee were certain that the column of redcoats had withdrawn out of fear.[13]

The real losers, however, were the 200 militiamen under the command of Captain Paul Demeré who were now isolated and besieged at Fort Loudoun. Because the distance to the fort was shorter from Virginia than Charleston, South Carolina appealed to Virginia for assistance; and a relief column of 300 men under the command of Colonel William Byrd started for Fort Loudoun in July. But Fort Loudoun's garrison was starving and ready to desert long before it drew near.

Captain Demeré was forced to ask for terms and surrendered the garrison on August 8, 1760. His soldiers and their families were to be escorted to Fort Prince George, but on the second day of the march, the Cherokee escorts attacked the column. Demeré was badly wounded, thirty-two others were killed, and the survivors were divided up as captives. Later, Demeré appears to have been tortured to death. Of the officers, only one, Captain John Stuart, was spared, because he was a former trader and a friend of Attakullakulla's.

Once more, South Carolina appealed to Amherst for assistance, and 1,200 British regulars returned to South Carolina, disembarking in Charleston on January 6, 1761. This time, James Grant was in full command as a lieutenant colonel. South Carolina also raised a regiment of its own. This included a company of provincials in which a farmer from Upper Saint John Parish named Francis Marion served as first lieutenant.

Grant's expeditionary force of some 2,800 regulars, provincials, and Mohawk and Stockbridge Indians reached Fort Prince George in May and found things calmer. Its commander had managed to ransom 113 white prisoners, many from Fort Loudoun, and win the respect of the Cherokee in the process. Grant might have stopped then and there and bargained for peace, but Amherst's orders were "to chastise the Cherokees [and] reduce them to the absolute necessity of suing for pardon." Revenge for the massacre at Fort Loudoun—which itself had been revenge for the slaughter of the Cherokee hostages at Fort Prince George—gave local fervor to Amherst's charge.

Moving northward from Fort Prince George, Grant's column was met by a force of about 1,000 Cherokee on June 10, 1761, just short of Echoe, where Montgomery had called it quits the year before. Once again, Cherokee warriors lay in ambush in a narrow pass and Francis Marion's company of South Carolina rangers was given the unenviable task of clearing the way. Despite heavy casualties, Marion advanced steadily, and Grant's main force soon followed. Routing the main body of Cherokee, Grant proceeded to carry out a systematic advance of death and destruction. All fifteen of the Middle Town villages were burned, along with 1,500 acres of cornfields and bean fields. Such widespread destruction would not be felt in the south again until Sherman marched from Atlanta to the sea a century later.[14]

With no assistance likely from the French at Fort Toulouse and with William Byrd's expedition from Virginia building a road into the northern part of their territory, the remaining Cherokee sent Attakullakulla to Grant to broker a peace.

Attakullakulla proved an astute negotiator. The terms that ensued pushed the Cherokee boundary westward, but only a small distance. Lyttelton's demands for retribution against those who had killed settlers in 1759 were quietly forgotten. Under the guidance of John Stuart, who was determined that Lyttelton's callous treatment of the Cherokee not be repeated, a measure of trade resumed between the Cherokee and Carolinians.

General Amherst recognized that such trade was necessary;

but in the aftermath of the Cherokee war, faced with similar issues on the northern frontier, he made an almost unilateral decision as commander in chief that was to have dire consequences. Forgetting, or at least choosing to ignore, the pleas of the dying John Forbes, Amherst viewed the Indian situation in strictly military terms and now that the French were removed from the situation, dictated that the long-established custom of dispersing presents was to cease.

"You are sensible how averse I am, to purchasing the good behavior of Indians, by presents," wrote Amherst to Sir William Johnson, "the more they get the more they ask, and yet are never satisfied; therefore as a trade is now opened for them, and that you will put it under such regulations as to prevent their being imposed upon, I think it much better to avoid all presents in the future." It was to be an ominous pronouncement.[15]

Meanwhile, as Francis Marion was honing the ranger skills that would serve him well as the "Swamp Fox" in a coming war, the most famous ranger of the current conflict had been dispatched westward to accept the French surrender of Detroit and see to French outposts as far away as the Straits of Mackinac. Major Robert Rogers had been engaged with Brigadier Haviland's push north from Lake Champlain; but barely was the ink of Vaudreuil's signature dry on the surrender documents at Montreal than Amherst sent Rogers west. "The capitulation including in it, as we hear," reported the *Boston Evening Post* of October 13, 1760, "not only all Canada, but likewise all the territories thereof depending—and that *Major Rogers with a large body of rangers were gone upon a distant expedition towards Lake Superior*."[16]

With two ranger companies, Rogers ascended the Saint Lawrence from Montreal, camped on the ruins of Fort Frontenac, paddled along Lake Ontario's northern shore to Niagara, and then portaged south into Lake Erie. As the main body of his rangers rowed along Lake Erie's stormy shores, Rogers pushed on ahead with a few others and detoured south from Presque Isle to Fort

Pitt, following Amherst's orders. Here, Rogers conferred with Brigadier Robert Monckton and was joined by the veteran Indian agent and trader George Croghan and a company of Royal American regulars for the push westward to Fort Detroit.

After regrouping at Presque Isle, the combined party of Rogers, Croghan, regulars, and rangers left there on November 4, 1760, in nineteen assorted whaleboats and bateaux. There was considerable uneasiness over how both the French at Fort Detroit and the Ottawa and Huron along the way would receive them. Croghan dispatched Iroquois and Delaware emissaries to invite Ottawa chiefs to meet them at the mouth of the Detroit River, but some thirty Ottawa signaled the flotilla to parley near present-day Ashtabula, Ohio, and others did so from the mouth of the Cuyahoga River. It was true, Rogers assured them; the French were no longer masters of Canada, and henceforth they must look to the British.[17]

More meetings with representatives of both the Ottawa and the Huron occurred as Rogers and his party moved westward past Cedar Point and Maumee Bay. No warfare broke out, however, and on November 27, 1760, the command entered the mouth of the Detroit River. Now, it was the French commandant at Fort Detroit who questioned their advance. Despite Rogers's letter containing the terms of Vaudreuil's surrender, the commandant expressed surprise that "no French officer accompanies you." But this matter, too, was resolved peaceably, and two days later Rogers's detachment of green-clad rangers stood beside the red coats and blue breeches of the Royal American regulars and watched as the French flag with its golden fleur-de-lis was lowered from Fort Detroit. It had flown there for fifty-nine years.[18]

But Major Rogers had not yet completed his full assignment. On December 7, 1760, a party of rangers set off to take over French posts southwest of Detroit. A day later, with thirty-five rangers and a dozen French inhabitants and Indian guides, Rogers himself set off for the Straits of Mackinac to enforce the surrender of Fort Michilimackinac. Old-timers marveled at his courage, but had their doubts. "Everybody here," wrote Captain Donald Campbell of the Royal Americans, "says he will find great difficulty to get himself to Michilimackinac even with a small detachment; they doubt even it's possible to be done."

Indeed, that proved to be the case. A week later, about seventy-five miles north of Detroit on the western shores of Lake Huron, winter got the best of the man who, just the year before, had made his epic return from Saint Francis. In the face of great ice floes and a hammering north wind, Rogers was forced to halt. "To our great mortification," the major wrote, "we were obliged to commence a return; in which, we were so much obstructed by ice, that we did not reach Detroit until the 21st."[19]

But still, Rogers was not content to rest. After arriving back in Detroit, he departed again on December 23, bound this time to report to General Amherst. By the time Rogers presented himself before Amherst in New York City on February 14, 1761—by his own admission "in perfect health"—he had made an incredible six-month circuit through parts still largely unknown to the British, heightening his already enlarged reputation.

Unfortunately, however, the dark side that always plagued Rogers had been present on this journey, too. Impoverished more than usual by his years of service around Lake George, Rogers entered into a trading partnership at Fort Niagara. Supplies were purchased from the partnership for the expedition, and other partnership goods became part of the expedition's cargo. This was not exactly illegal—indeed, John Bradstreet had taken similar liberties—but it was what in a later time would be called an "appearance of impropriety." In some respects, the entire situation was a double-edged sword. Were it not for the entrepreneurship of their officers, many small units would have gone without supplies. But such arrangements did provide an easy mark when critics chose to cry "foul."

Then, too, there was the deed. Rogers had been eager to claim lands since his early days on the New Hampshire frontier. Now, he returned to New York with a deed from four Chippewa chiefs purporting to convey an estimated 20,000 acres of land on the southern shores of Lake Superior. The acreage lay at the extreme western end of what would become Michigan's Upper Peninsula and even then was rumored to hold rich copper deposits. It was not the sort of thing that British officers routinely acquired—no matter how accustomed they might be to the spoils of war—and it would come back to haunt him.[20]

Regardless of Robert Rogers's indiscretions and his failure to reach Fort Michilimackinac that fall, the very fact that he had been bound there to accept its surrender was proof in itself that Great Britain had indeed taken all of Canada. Whether Great Britain would choose to keep its North American conquests would soon be a matter of great debate. Meanwhile, however, there were still other battles raging in Mr. Pitt's global war.

17

MARTINIQUE TO MANILA

Since so many of Europe's rivalries had been born of the issue of royal succession, it should have come as no surprise that the passing of Ferdinand VI of Spain in 1759 sparked another. Ferdinand had stoutly and repeatedly refused to engage Spain on the side of France. To be sure, he had his disagreements with Great Britain—among them its blatant harvesting of timber from Spanish Honduras while denying Spain fishing rights in the Grand Banks of Newfoundland—but these paled beside the prospect of all-out war. Ferdinand's successor, his half brother Charles, son of Philip V by Philip's second marriage and yet another great-grandson of Louis XIV, felt quite differently. Having ruled Naples as Charles III since 1735, he was determined to support his Bourbon kin in France and revitalize Spain's role in world affairs.

It didn't help Charles's clarity of thought that he despised Great Britain because of a situation stemming more from personal ego than geopolitics. As king of Naples, he had suffered humiliation when a British squadron suddenly appeared in the Bay of Naples during the War of the Austrian Succession and its commodore allowed him but one hour to withdraw his Neapolitan troops from acting in concert with those of Spain. Now, even before Charles reached Madrid to assume the Spanish throne, he was sending missives throughout Europe. On the one hand, he offered to serve as a neutral mediator between Great Britain and France; but on the other, he sought British concessions on the issues of logging and fishing, as well as the seizure of certain Spanish ships.

Charles's personal diplomacy culminated in August 1761, when the ruling Bourbons of Spain, France, Naples, and Parma signed the Bourbon Family Compact, essentially pledging that the enemy of one of its members automatically became the enemy of the others. Specifically, the compact provided that should France and Great Britain not make peace by May 1, 1762, on terms that also resolved Spain's grievances, Spain would declare war on Great Britain. Charles III had taken the long neutrality of Ferdinand VI—which had actually been an aid to Great Britain in the early years of the war as Pitt built up its navy—and pushed Spain toward France until Spain now irrevocably cast its lot with Louis XV. It would prove too little, too late for France and would show why some in Spain had called Charles III's predecessor, Ferdinand, "the Sage."[1]

Such machinations, of course, did not escape the eyes and ears of William Pitt. War with Spain? Bring it on! Far from being cowed by the prospect, Pitt relished it and unrolled maps of the Caribbean and far Pacific. While the Royal Navy was busy gathering the plums of France's colonial empire for Britannia, it might just as well also gather some of Spain's. Not all in Great Britain felt this way, of course, in part because in Britain, too, there had been a change of monarchs. George II died on October 25, 1760, struck down suddenly by an apparent heart attack at the age of seventy-seven. The old king had come to revel in the victories that Pitt's vision had wrought, and there was no early indication that his successor, his grandson, would do otherwise.

But George III was different. "Born and educated in this country, I glory in the name of Briton," declared the twenty-two-year-old in his first address to Parliament, in a clear attempt to disassociate himself from his ancestors' fixations with Hanover. But young George III exasperated Pitt and raised even Newcastle's eyebrows when he sought to characterize the war that was winning so much empire as "bloody and expensive." The new monarch urged a speedy termination, albeit with "an honorable and lasting peace."

Pitt looked around for the puppeteer holding the royal strings and found him in John Stuart, the earl of Bute. Having lost his own father at thirteen, George III seems to have idealized this older Scot, who whispered in the royal ear all the appropriate reassurances, while pushing his own plans. Clearly, those included stripping Pitt of his powers as de facto prime minister.

Interestingly, at the moment of his ascension to the throne, George III was also in the process of choosing a wife. Had he truly meant to make good on his promise of glorying "in the name of Briton," he might have chosen Lady Sarah Lennox. She was the daughter of the duke of Richmond, as charming, as blue-blooded, and as English as anyone might desire, and her family ties ran throughout the upper echelons of Pitt's Whig party. By all accounts, the king was smitten by her and she with him, but he bungled the courtship with sophomoric prattle.

The king's mother, the dowager princess Augusta, and Lord Bute took advantage of these stumbles to intervene and quash a union that no doubt would have strengthened Pitt's hand and brought George III out of his shell. On July 8, 1761, the young ruler, who had said that henceforth he would look to Britain, announced with apparent unease to Pitt and the cabinet that he would marry Princess Charlotte of Mecklenburg-Strelitz. His mother and Bute had won, and had assured themselves that the king would continue to be under their influence to the detriment of Pitt.[2]

The pending war with Spain was the paramount issue that brought about Pitt's downfall. Confident that he alone was the architect of the expanding empire, Pitt pushed for an immediate declaration of war against Spain at a cabinet meeting on September 18, 1761. With Lord Bute no doubt keeping score, only Pitt's brother-in-law, Lord Temple, supported him. Having relied so long on Newcastle's power in the House of Commons to bolster his aims, Pitt was now very much adrift.

At a subsequent cabinet meeting on October 2, he held firm to his position that war must be declared immediately against Spain or he would resign. Still, he could change no minds; and two days later, William Pitt, the man who had once boasted that "I can save England" and indeed had done so, tendered his seals of office to George III. Significantly, the king made no attempt to dissuade him.[3]

Two centuries later, after another but far more deadly global war, it became popular to compare William Pitt to another English bulldog, Winston Churchill. Both had seen the global picture, and each had led Great Britain back from a brink. Both, too, were repudiated by their fellow countrymen at the moment of victory.

A month later, a reshuffled British cabinet "asked Spain to disavow its suspected alliance with France." When Charles III refused, Great Britain declared war on Spain on January 4, 1762. Perhaps Pitt had been right after all, especially after Spain showed its true colors and quickly moved to conquer its neighbor, Portugal, heretofore a neutral country friendly to Great Britain. Should Portugal fall, its great colony of Brazil and all of its lucrative trade would come to Spain.

Who should the British dispatch to Portugal's aid but Lord Loudoun and Brigadier General John Burgoyne? The former had already come to grief in North America, though the latter's day of reckoning at Saratoga was still fifteen years away. With 6,000 troops outside Lisbon in the summer of 1762, Loudoun and Burgoyne stymied Charles III's drive on the Portuguese capital. Spain's failure to conquer Portugal should have given Charles pause as to where his warrior policies were leading Spain.[4]

Even with Pitt out of office, his reach continued to be felt by expeditions already under way. Some battles had been fought and won. India was a case in point. In January 1757, Robert Clive had led a force to recover Calcutta from the nawab of Bengal and crush French influence there. Then, learning of the official declaration of war between Great Britain and France long after the fact, Clive led 3,000 men, mostly trained sepoys from Madras, against the nawab's forces at the battle of Plassey on June 23, 1757. After Clive's victory at

Plassey, both the British and the French dispatched reinforcements and determined to fight for the subcontinent and its lucrative trade.

For a time, the British dominated around Clive's movements in Bengal in the north and the French extended their influence from Hyderabad along the Coromandel Coast to the south. Here too, however, the ultimate disposition was determined by Great Britain's mounting naval superiority. In the spring of 1758, the French Admiral Anne Antoine, comte d'Aché, arrived at Pondicherry with a squadron numbering about eight ships of the line. Aided by this fleet, French land forces captured Fort Saint David south of Pondicherry and laid siege to the British base at Madras. Then, a comparable English squadron under the command of Rear Admiral Sir George Pocock not only lifted the siege of Madras but also blockaded the French at Pondicherry.

In time, it didn't matter how strong the French were on land, because they couldn't obtain fresh troops and supplies via the sea. What's more, the British controlled an ever-increasing portion of trade in the Indian subcontinent, which was the reason to have these colonies in the first place. D'Aché's and Pocock's squadrons fought skirmishes and three major engagements—the final one on September 10, 1759, in the Bay of Bengal south of Pondicherry—but the British routinely got reinforcements and the French did not.

After the French were defeated in Quiberon Bay, no reinforcements of any kind reached India. On January 20, 1760, British land forces decisively defeated the French at Wandiwash north of Pondicherry and proceeded to capture French trading stations up and down the Coromandel Coast. By April 16, 1760, only Pondicherry remained in French hands in eastern India. It endured a grisly siege before finally surrendering the following January. The French surrendered their last stronghold, Mahé on the west coast, a month later; and, for better or for worse, India began almost two centuries of servitude as a British colony.[5]

Pitt's further plans for the West Indies were also well under way by the time he left office. Hopson had attempted to seize

Martinique in 1759 and had settled for Guadeloupe, but Martinique still remained a prize to be sought. As the key base for French privateers in the Caribbean, it was much more than just another sugar island. France recognized this and, despite dwindling resources, managed to reinforce the garrison at Fort Royal with 250 additional troops in 1760 and another 500 the following year. The newly arrived governor general of the French West Indies, Le Vasser de la Touche, boasted that Martinique's militia numbered 20,000; but most of this force seems to have existed in name only, and indeed the new governor assembled the militia but once before the British invasion.

Pitt had delayed the sailing of the Martinique expedition until after the hurricane season, but in the meantime had ordered Amherst to send 2,000 men from North America to attack Dominica just to the south of Guadeloupe. Dominica was, of course, one of the so-called neutral islands, but such characterizations were long past anyone's consideration. Interestingly, though, Dominica was one of the few refuges of the indigenous Caribs who had been extirpated from most of the other islands of the Lesser Antilles. By 1761, some 2,000 French settlers were living on the island along with the Caribs, but it was not garrisoned by French troops.

A tiny flotilla of four warships and four transports carrying about 700 regulars, including some of Montgomery's Highlanders, anchored off the principal town of Roseau on June 6, 1761, and moved ashore to occupy the town. When militiamen on the surrounding hillsides began a steady small-arms fire, the grenadiers advanced rapidly up the slopes and dispersed them. By nightfall, in a rapid about-face, the French inhabitants were streaming into Roseau delivering up their arms and begging to be given the same favorable terms as the inhabitants of Guadeloupe.

If the force thrown against Dominica seemed small, Pitt was determined to have more than enough troops and ships to complete the task at hand on Martinique. To command the Martinique expeditionary force, Amherst chose Robert Monckton, who was now recovered from his wounds on the Plains of Abraham and had been rewarded for his role there with both a major generalship and an appointment as governor of New York. Monckton would command 14,000 troops drawn from North America, Europe, and the Caribbean.

The naval force was to be equally stout. Its commander was Rear Admiral George Rodney who had served under Boscawen against Louisbourg and had frustrated Amherst by interrupting the general's Atlantic passage in 1758 to engage a French ship. Rodney sailed from England in early October 1761 and rendezvoused in Barbados with the transports from America bearing Monckton's force. In all, his fleet consisted of eighteen ships of the line, fourteen frigates, nine sloops and bomb ketches, and some 11,000 men. It was another example of the vast resources of the Royal Navy.

On January 5, 1762, Monckton and Rodney with an invasion fleet of 173 vessels weighed anchor in Carlisle Bay in Barbados and sailed for Martinique. With Monckton as his brigadiers were two veterans of the campaigns against Canada: William Haviland and James Grant. Their troops made three attempts to force a landing in and around Fort Royal Bay and then on January 16, 1762, finally landed west of the fort where Hopson had come ashore two years before.

Once again, French defenders fired on the beachhead from the surrounding hillsides; but unlike Hopson, Monckton simply pushed more troops and artillery onto the beaches, and a week later Grant's grenadiers stormed these heights and others to the east above Fort Royal. This was enough for de la Touche's vaunted militiamen and they dispersed into the hills while his regulars crowded into the fort's citadel. Monckton hauled artillery to heights overlooking the fort and began a bombardment that lasted until the garrison of 800 surrendered on February 3, 1762.

Governor General de la Touche escaped the siege and retreated north to Saint Pierre in an attempt to rally resistance. What he found was similar to the situation confronting the French army when it had counterattacked in the final days of the

Guadeloupe campaign two years before. The French planters simply had no stomach to oppose the British and suffer losses to their property, when by capitulating they could very likely broker the same favorable terms extended to Guadeloupe. De la Touche recognized the inevitable and surrendered the entire island. Monckton's troops peacefully entered Saint Pierre on February 16, 1762.

With the fall of Martinique, the resolve of the French in the West Indies evaporated. Rodney sent squadrons to the islands of Saint Lucia, Grenada, and Saint Vincent, and all surrendered without firing a shot. Representatives from these islands met with Monckton shortly thereafter and procured favorable terms. By all accounts, most French inhabitants of the islands were delighted. Having watched their produce sit bottled up on wharves or seized by English or American privateers, they were now accorded the security and profits that came from trading with the world's greatest economic force.[6]

Meanwhile, Great Britain looked around the Caribbean for the treasures of its new enemy, Spain. There was none greater than Havana, Cuba. For two centuries, Havana had been the military and commercial hub of Spain's Caribbean colonies and the gateway to its conquests on the mainland from Saint Augustine to Cartagena. Havana, commanding city of 35,000, boasted a fine deepwater harbor, guarded on the east by Morro Castle and on the west by the Punta. Great Britain had tried to capture it during the War of Jenkins' Ear, and now Charles III seemed to be foolishly offering it up.

In Pitt's absence, First Lord of the Admiralty Anson and Lord Ligonier issued orders against Havana only three days after the declaration of war, and an expeditionary force sailed from Portsmouth on March 6, 1762. These forces rendezvoused with those of Monckton and Rodney recently released by the fall of Martinique.

But France and Spain had one last chance. A French admiral, the comte de Blénac, broke loose of the Brest blockade and arrived in the West Indies with what was arguably the last French squadron of any size still at sea—about eight ships of the line.

Blénac was too late to save Martinique; but by joining forces with a Spanish fleet from Havana, he might be able to surprise a British squadron or threaten British Jamaica.

The governor of Jamaica was in fact so alarmed at the threat that he insisted on having British ships remain in the harbor at Kingston rather than letting them move to intercept Blénac as the British commodore proposed. But cooperation between the French and the Spanish, which seemed so easily mandated by the Bourbon Family Compact, broke down. The Spanish fleet of twelve ships of the line never sortied from Havana. Blénac was forced to seek refuge at Cap François on Santa Domingo, the last French-held island in the Caribbean, and ended up blockaded there.

That left Havana wide open. On June 7, 1762, British troops landed about six miles east of town and proceeded to lay siege to Morro Castle. A similar force landed west of the city under the command of Colonel William Howe, still adding to his colonial military experience. But as the British siege line extended and artillery began to bomb the town, Havana's most formidable defense showed itself. It was midsummer, and a host of tropical diseases—yellow fever, malaria, and gastrointestinal disorders known and unknown—ravaged the British troops. Havana might fall, but would there be any British troops left standing to march through its gates?

Relief came in the form of the arrival of fresh troops from North America, half of them provincials from New York, New Jersey, Rhode Island, and Connecticut. On July 30, a mine was tunneled under Morro Castle. It was exploded and then the fort was stormed. The Punta was taken out of action on August 11; and two days later—just in the nick of time from the British perspective—the city surrendered.

At the cost of an estimated 7,000 dead—most from disease—Great Britain had secured the cornerstone of Spain's empire in the New World. It had also captured 3 million pounds in gold and silver, twelve ships of the line, and several frigates—one-quarter of the entire Spanish navy. Henceforth, whatever treasure ships dared to try to reach Spain from Mexico would be forced to assemble far to the south at Cartagena. When Charles III heard of this, his youthful indignation at the British fleet off Naples must have paled by comparison.[7]

With the Caribbean almost an English lake and the subcontinent coming to heel, a spin of the globe revealed one more tempting target. As First Lord of the Admiralty Anson and Lord Ligonier listened, Lieutenant Colonel William Draper, a veteran of the victory at Wandiwash, described a plan to attack Manila, the commercial and political hub of the Spanish Philippines. Manila was six to eight months by sea from England, but Draper assured Anson and Ligonier that all the necessary forces for the expedition were already in India, only six to eight *weeks* away. In fact, given the lengthy delays in communications between Madrid and Manila via Mexico, if it moved quickly, a British force might just be able to appear off Manila before the garrison there even learned of the declaration of war.

Anson and Ligonier agreed to the plan, and Draper left England in February 1762 with a temporary commission as a brigadier general and authority to organize an expeditionary force in Madras. Field commanders in India were less enthusiastic, however, and reluctant to part with troops. Draper managed to augment his own Seventy-ninth Foot Regiment and a company of royal artillery—almost 1,000 regulars—with 600 sepoys, two companies of French deserters, and several hundred assorted locals. "Such a banditti never assembled since the time of Spartacus," Draper grumbled as his little flotilla of fifteen ships finally departed Madras at the end of July.

As the British ships sailed into Manila Bay on September 22, 1762, the silence of the guns at its guardian fortress of Cavite proved that Draper had indeed caught the Spanish unawares. Lacking the troops to encircle the town in a protracted siege, Draper subjected Manila to a heavy artillery barrage and then stormed a portion of its walls before its garrison capitulated on

October 6. By the end of the month, Spain formally surrendered all of the Philippines to Great Britain.

At first glance, Great Britain's capture of Manila was the ultimate global grab, a triumph that demonstrated the might of the British army and the Royal Navy halfway around the world from London. There was, however, one element of this military operation quite different from those against Canada, India, and the sugar islands of the West Indies. Draper may have captured Manila, but he had hardly conquered the Philippines. These islands and their inhabitants did not join the trading frenzy of the British Empire. Instead, certain Spanish colonial officials mustered a Filipino guerrilla army of 10,000 natives—three-quarters of them armed only with bows and arrows—and limited British control to Manila and Cavite. It was a situation destined to be repeated 136 years later after an American naval squadron under Admiral George Dewey steamed into Manila Bay at the height of the short-lived Spanish-American War. This global reach of a rising American imperialism forced Spain to surrender the islands again, but spawned a similar guerrilla insurgency that went on for decades.

When the Philippines were finally returned to Spain on May 31, 1764, under the terms of the Treaty of Paris, the British appeared to be surrendering Manila far more than merely withdrawing from it. The lesson, according to the historian Fred Anderson, was that "only the voluntary allegiance, or at least the acquiescence," of colonists could maintain an empire and that "when colonial populations that refused their allegiance also declined to trade, the empire's dominion extended not a yard beyond the range of its cannons." This lesson of the Philippines went unheeded, however, both by the United States more than a century later and by Great Britain itself in regard to its North American colonies barely a few years hence. There definitely were limitations to empire.[8]

18

SCRATCH OF A PEN

When Vaudreuil signed the capitulation of Montreal and with it all of Canada, France was expelled from North America east of the Mississippi. As Francis Parkman so succinctly put it, "half a continent had changed hands at the scratch of a pen." And Louis XV had fared no better on global battlefields. In Europe, Africa, India, and throughout the Caribbean—primarily because of the sea power that William Pitt had championed—Great Britain had picked one jewel after another from the French crown. Indeed, as the naval historian Alfred Thayer Mahan pronounced, with ample evidence, "at the end of seven years, the kingdom of Great Britain had become the British Empire."[1]

But how—and even if—all these vast territorial gains were to be consolidated by Great Britain was a matter of great debate. Trading captured territories at the peace table was a time-honored ritual of European diplomacy. France had grabbed Minorca early in the current conflict in expectation of just such a round of trading. By 1762, a look at the globe showed that Minorca was about all that France had left to trade, but this did not keep the French minister of foreign affairs, Choiseul, from bartering for all he was worth. As it turned out, Choiseul's best allies in the process proved to be Lord Bute and those in Great Britain who simply could not fathom the scope of Pitt's victories. Surely, their argument ran, a return of at least some of France's captured territories was essential to restoring some semblance of a balance of power in order to avoid another conflict.

While Pitt was still in power, he eschewed such talk of trades, but the great debate in England after Vaudreuil's surrender nonetheless became whether Great Britain should keep Canada or Guadeloupe. Would it be land or sugar, land or trade? A flurry of pamphlets appeared on the subject, including one written anonymously, but not very secretively, by Benjamin Franklin: "The Interest of Great Britain, Considered with Regard to Her Colonies, and the Acquisitions of Canada and Guadeloupe." Franklin argued for keeping Canada. Warming to his own reasoning, he later wrote that not only did "the foundations of the future grandeur and stability of the British Empire lie in America," but also "if we keep [Canada], all the country from Saint Lawrence to Mississippi will in another century be filled with British people."[2]

But the immediate financial rewards certainly seemed to argue for keeping Guadeloupe. Its trade and potential for tax revenues dwarfed Canada's. Because of this, during the initial bargaining, Choiseul seemed as determined to keep Guadeloupe as Pitt was determined to acquire Canada. Those in Britain who argued for Canada, looking through less visionary glasses than Franklin's, nonetheless saw owning it as a way to rid their North American colonies of a meddling neighbor once and for all. But then Dominica fell, and in February 1762—with Pitt now out of office—Martinique also fell. Suddenly the question was no longer Canada or Guadeloupe, but Canada or all of the French West Indies.[3]

Choiseul, who had finally managed to rise above the muddle of the French bureaucracy and consolidate the ministries of foreign affairs, war, and the marine into one portfolio, was all too pleased to negotiate with Bute rather than Pitt on this issue. Too late, France found the one firm voice that Great Britain had once had in Pitt, and Choiseul did his utmost to salvage a workable peace out of the disasters of the last few years. Not the least of his problems was his Spanish ally. "Had I known what I now know," Choiseul sighed on the dismal failure of the Spanish war machine everywhere, "I should have been very careful to cause to enter the war a power which by its feebleness can only ruin and destroy France." While Charles III of Spain still demanded British concessions, Choiseul made it his priority to end the conflict and save what was left of France.[4]

But France and Spain weren't the only allies whose ties had been frayed by the course of the war. Great Britain had originally joined forces with Frederick the Great to support the balance of power in Europe generally, and King George II's Hanoverian roots specifically. As to the latter, even without an English wife, George III was seeking to cut his ties to the old familial homeland and thereby save the annual subsidy to Frederick.

Lord Bute urged the young monarch in this direction and believed that since France had lost its worldwide empire, there was no longer any great advantage for Great Britain to distract France on the continent by continuing to support Frederick. On April 30, 1762, Bute's cabinet narrowly voted to discontinue both the continental war and the Prussian subsidy, essentially clearing the way to make an independent peace with France. "The military and diplomatic partnership of Great Britain and Prussia was thus prematurely dissolved before the war had run its full course."[5]

Somehow, Frederick the Great managed to survive what he considered a clear desertion by Bute by once again proving that, if nothing else, he had enormous staying power. Frederick got considerable assistance in this regard by the fortuitous death of Empress Elizabeth of Russia, one of the three European women who had been once arrayed against him. Elizabeth's successor, young Peter III, immediately hailed Frederick as his "lord and master" and signed a peace treaty with Prussia that not only restored all of Russia's territorial conquests, but also placed the Russian army at Frederick's disposal.

Peter's turnaround in foreign policy offended many in Russia, not least his wife, Catherine, who quickly acquiesced in her husband's murder and ascended the Russian throne herself in June 1762, destined to become Catherine the Great. She deemed her future greatness to lie within Russia and was not interested in the renewal of a war against Prussia—and certainly not in a war against Austria for Prussian interests.

So, with France bloodied enough on the continent for one generation, Choiseul at last consolidated the power necessary to override the marquise de Pompadour's influence, which had drawn France into this mess in the first place. Choiseul told Bute that France, too, would withdraw its troops from central Europe. This left Maria Theresa of Austria to face Prussia alone over the issue of Silesia that had simmered for almost a quarter of a century.[6]

On November 29, 1762, the final draft of the Treaty of Paris, principally between Great Britain and France, with Spain participating reluctantly, was read before both houses of the British Parliament. Its sweeping geographic terms underscored the global war it sought to end. All of Canada and all French claims east of the Mississippi were ceded to Great Britain, except the city of New Orleans. This presence aside, British subjects were guaranteed free rights of navigation on the entire length of the Mississippi. French fishing rights to cod in the Grand Banks, originally granted by the Treaty of Utrecht, were confirmed, and the two tiny islands of Saint Pierre and Miquelon off the southern coast of Newfoundland were given to France as commercial bases. Louisbourg, its military fortifications destroyed in 1760 under orders from Pitt, was to remain British.

In the Caribbean, three of the so-called neutral islands—Saint Vincent, Dominica, and Tobago—along with Grenada and the Grenadines were also to remain British. France, however, was restored to the possession not only of Guadeloupe but also of Martinique and Saint Lucia. (Land in North America and the security it seemed to promise the North American colonies had won out over sugar and trade.) In Africa, Great Britain retained Senegal, but the far more important trading station of Gorée was returned to France. In India, France was limited in its future activities to pre-1749 stations for trading purposes only. France also restored to Great Britain two trading posts on Sumatra.

In Europe, the main bargaining chip that France still held, Minorca, was returned to Great Britain in exchange for the naval base at Belle-Ile-en-Mer off the French coast that Pitt's cross-Channel raids had finally captured. Last, both the British and the French withdrew from Hanover and from any military operations for or against Frederick.

And what of Spain? How had Charles III fared after his ill-advised rush to the Bourbon banner? Spain's belated allegiance to France cost it dearly. Spain came out on the down side of all of its grievances with Great Britain and then some. Spain was not to participate in the cod fisheries; it must permit British logging rights in Honduras, albeit without fortifications; and all issues of Spanish ships seized by Great Britain before the declaration of war were to be decided in British admiralty courts. Finally, there was the matter of Havana. Great Britain would return Havana and the island of Cuba to Spain, but only in exchange for all of Florida. This gave Great Britain control of the North American coastline from the icy limits of Newfoundland to New Orleans.

If Choiseul felt any remorse in forcing this last point on his weakling ally, he sought to appease Spain's loss of Florida by conveying New Orleans and all of France's remaining claims west of the Mississippi—a vague and unknown land called "Louisiana"—to Charles III. Choiseul considered this an empty gesture because Louisiana appeared almost worthless. Scarcely four decades later, however, Napoleon would bully Spain into returning the territory to France, so that France could in turn sell it to the young United States. The change in the perceived value of Louisiana occurred, in part, because, as Benjamin Franklin wrote prophetically in March 1763, "there appears everywhere an unaccountable penchant in all our people to migrate westward."[7]

The terms of the Treaty of Paris were debated in the House of Commons on December 9, 1762. It was not a constitutional requirement that Parliament approve the treaty, but Lord Bute considered it a prudent political move, if for no other reason than to diffuse a storm of popular criticism against both the treaty and him personally. Indeed, the reaction of the majority of the common people was that the great victory won by their ardent champion Pitt was now being frittered away by Bute, who esteemed it far too cheaply. As the historian Walter Dorn characterized it, the Treaty of Paris was "widely condemned in Great Britain and so generally regarded by British historians as falling in its terms far below what their conquests gave Englishmen a right to expect."[8]

Thus William Pitt was to have his valedictory. After some hours of debate that December day, the doors of the House of Commons opened to reveal the "great commoner" himself. Racked with pain, his gouty legs wrapped in heavy cloth, Pitt was carried to his seat, from which he proceeded to speak on the subject at hand for over three hours. His discourse was far-ranging, at times less than coherent, but the architect of empire summarized his vehement opposition in one sentence. The proposed treaty, Pitt declared, "obscured all the glories of the war, surrendered the dearest interests of the nation, and sacrificed the public faith by an abandonment of our allies." Never had he, nor would he have, Pitt further avowed, "made a sacrifice of any conquest."

Speaking in opposition to Pitt and in support of the treaty, the young earl of Shelburne told the House of Lords that Great Britain had not entered the war to conquer the territories of other nations. Its goal, Shelburne maintained, had been simply "the security of our colonies upon the continent of North America, threatened by French encroachments." That issue having now been resolved and Great Britain "having made very large demands in North America, it was necessary to relax in other parts of the world."[9] How perceptions had changed!

Even the battered old duke of Newcastle, who had long been Pitt's reluctant partner and who had finally resigned from Bute's government in disgust, hoped that somehow Pitt would arouse enough opposition in Parliament to force a stiffening of terms. But that was not to be. At the end of a long day of speeches, the House of Commons voted 319 to 64 in favor of the treaty. In the

House of Lords, the opposition was so weak that it did not even demand a division to the overwhelming voice vote in favor.

On February 10, 1763, the Treaty of Paris was officially signed. It was far different from what might have been brokered by William Pitt, and one could almost hear the sigh of relief escaping from the shrewd Choiseul. Five days later, Frederick the Great and Maria Theresa agreed to the separate Treaty of Hubertusburg between Prussia and Austria. As to all disputed territories between the two, it provided for a *status quo ante bellum*: each party would keep what it had in the beginning. In effect, Europe had finally worn itself out. During nine years of global conflict, the map of the world had changed considerably. The map of Europe had remained remarkably the same.[10]

Despite the global actions and the European feuds that had precipitated them, the greatest effect of the Seven Years' War—the war that North Americans would call the French and Indian War and date as nine years in length, not seven—was to deliver Canada into the hands of Great Britain. The debate had once been Canada or Guadeloupe, but in retrospect it poses an interesting "what if?" Had Great Britain retained Guadeloupe and Martinique instead of Canada, might the trade from the sugar islands have paid for a sizable chunk of Great Britain's war debt, thus avoiding the heavy taxation that would soon rile its North American colonies?

In many respects, these colonies had given their hearts and souls, as well as their purses, to His Majesty's war efforts. The Carolinas and Georgia had always been more concerned with Indian and Spanish issues along their borders; others had frequently seemed aloof; but Massachusetts, New Hampshire, Connecticut, and New York in particular, with Pennsylvania and even Virginia close behind, had strongly supported the war effort. "Nothing," said Governor Thomas Pownall of Massachusetts at the war's closure, "can eradicate from the [English] colonists' hearts their natural, almost mechanical affection to Great Britain."[11]

19

A Matter Unresolved

As important as the Treaty of Paris was to restoring world order, another document emerged from the year 1763 that was arguably to have an even greater impact on the future of the British Empire, particularly its North American colonies. On July 4, 1763—once again the date was to prove an ironic coincidence—instructions were sent to all colonial governors in North America forbidding settlements west of the crest of the Appalachian Mountains. Under penalty of dismissal, the governors were further restrained from making any grants of such lands. Three months later, these instructions were incorporated into what came to be called the Proclamation of 1763. In the end, about all the Proclamation of 1763 would prove was that in Great Britain's rush to bring order to its newly won empire, it had left one thorny matter unresolved.

Ostensibly, the Proclamation of 1763 was calculated to reduce friction with Indians in those territories recently won from France. In reality, however, it failed to appease the Indians and also served to frustrate westward-looking colonists from New York to the Carolinas. The royal decree created three new colonies: East and West Florida, just won from Spain; and what Great Britain would call Quebec, essentially all of New France north of Lake Champlain and downstream on the Saint Lawrence from near Montreal.

All land between these new colonies on the north and south,

the crest of the Appalachian Mountains on the east, and the Mississippi River on the west—including the hard-won Great Lakes and Ohio River valley—was designated a vast Indian reserve. British forts would be maintained in this territory and transient traders allowed, but permanent settlement was prohibited. Any British subjects then living west of the Appalachians were to "remove themselves from such settlements."

That, of course, was not going to happen. It was utterly unthinkable to most colonials that British regulars who had just won New France would now be used to keep them out of territory they had coveted since George Washington walked into Fort Le Boeuf. But before any evictions could be undertaken—indeed, even before news of the proclamation reached the Ohio River country—a host of Indian nations had taken matters into their own hands to oust their new British landlords.

And "landlords," it seemed to the Indians, was only too true and too distasteful a characterization. What had happened, the Delaware and other nations wondered, to the promises of the Treaty of Easton? Not only were the British fortifying old French posts such as Fort Detroit, but they were also building new ones. Most alarming of all was the permanent community rapidly rising in the shadow of Fort Pitt. Something had to be done.[1]

It was Francis Parkman who originally characterized the Indian attacks of the summer of 1763 as the "conspiracy of Pontiac." Indeed, such was the title of Parkman's volume on the subject, penned more than a century after the actual events. Subsequent historians embraced this characterization and embedded it in a chain of Native American unrest running from Roanoke to Wounded Knee. How much the Ottawa chief Pontiac was the central player—as opposed to merely a key player—continues to be debated.

What does seem clear is that a Delaware prophet or holy man, Neolin, had a vision that called on all Indian nations to reject their dependence on European-Americans—French or British—through an avoidance of trade, a return to ritual warfare and diets, and the gradual abandonment of European-made goods. The Seneca, led

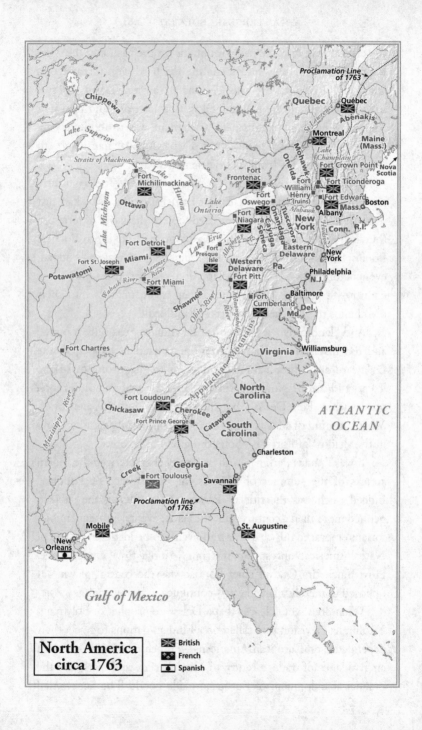

North America
circa 1763

⊠ British
☗ French
⊡ Spanish

by Kaiaghshota, took this vision one step farther and plotted for the outright ouster of their British landlords. In doing so, the Seneca broke with their brethren in the Six Nations who remained friendly to the British only because of the influence of Sir William Johnson. But the Seneca found allies in the Indian nations farther west that traditionally had been more closely allied with the French. Among these were the Ottawa, Huron, Potowatomie, and Chippewa.

The British manner of dealing with these nations of the Great Lakes country was far more arrogant and self-righteous than the more genial camaraderie of the French. The British took their right to occupy Indian territory as a matter of course. But the real incendiary factor proved to be Jeffery Amherst's decision to stop the long-established custom of presents. Announced in 1760, Amherst's policy of eliminating cash outlays for gifts was taken to the extreme after the French surrendered. The Ottawa, Huron, Potowatomie, and Chippewa around Detroit, "long accustomed to French finery for their women," were now cut off from any trade goods. Failing to heed the warnings of the dying John Forbes, Amherst stubbornly insisted that all Indians trade for their merchandise on the basis of a published schedule of prices.[2]

Enter Pontiac. Little is known of his early life, but learned speculation suggests that he was part Ottawa, and part Chippewa, and about forty years of age in 1763. His place of birth is also a mystery, although it may have been in the Ottawa village across the river from what was then the French outpost of Detroit. Certainly, Pontiac seems to have considered this area his home territory. It has long been suggested that Pontiac led Ottawa warriors from the Detroit area and participated in Braddock's Defeat, but even Parkman was careful not to state this as unequivocal fact. Ottawa warriors played a large role in the massacre at Fort William Henry, but again there is no hard evidence of Pontiac's presence there.

The first documented mention of Pontiac's activities has to do with a speech he gave at Fort Duquesne, probably in 1757. Recognized as a chief of the Ottawa, Pontiac reported the efforts of George Croghan to lure the western Indians to the British side and professed his loyalty to the French, no doubt in expectation of presents. Parkman wrote that Pontiac's first documented appearance was three years later when he greeted Major Rogers on the southern shores of Lake Erie as Rogers was headed west to receive the surrender of Fort Detroit. Rogers's original journal makes no mention of the encounter, and Parkman chose to recount the version given in Rogers's later writings, which seems to have been Rogers's embellishment after Pontiac had become well known.[3]

Whatever his earlier activities, by the fall of 1762 Pontiac was chafing under Amherst's "no gifts" policy and clearly held some measure of influence among the Ottawa, Huron, Potowatomie, and Chippewa in the vicinity of Detroit. Parkman's assertion that Pontiac "sent out ambassadors to the different nations" calling for concerted strikes against all British posts the following spring is, however, an inflation of his role. For one thing, the Seneca were already well into preparations for an attack against Fort Niagara on their own. Indeed, even among Pontiac's own Ottawa, there had been dissension, and the Ottawa whom Major Rogers encountered on Lake Erie were a pro-British faction that had abandoned the Detroit villages.

But it is also clear that Pontiac's zeal to defend both his homeland and the now lost French cause made him a powerful voice in the western Great Lakes country. His rhetoric, delivered by beads of wampum, probably urged attacks against British posts in western Ohio, Indiana, and southern Michigan in the spring of 1763. On May 16, tiny Fort Sandusky on Lake Erie became the first British post to fall when neighboring Huron captured its commander by deception and killed its fifteen-man garrison. The attackers also killed a number of merchants and made off with their trade goods. Garrisons of similar size at Fort Saint Joseph (present-day Niles, Michigan), Fort Miami (present-day Fort Wayne, Indiana), and Fort Ouiatenon (present-day Layfayette, Indiana, on the Wabash River) fell to similar attacks.

Then on June 2, a band of Chippewa—who, like the Seneca, appear to have been acting independently—staged an enthusiastic lacrosse game outside Fort Michilimackinac at the Straits of Mackinac. The heretofore friendly Chippewa whooped and hollered and made a great show. Soldiers and officers alike from the thirty-five-man garrison gathered outside the open fort gates to watch. As the game progressed, Chippewa squaws casually sauntered into the fort with weapons hidden under their blankets. When an apparently errant shot sent the leather ball sailing over the fort walls, the warriors ran through the gates and followed the play. Once inside, they grabbed the hidden weapons and proceeded to slaughter nearly the entire British garrison. French traders at the post were not harmed.

To the east, Delaware and Shawnee warriors attacked settlements on the Monongahela and later laid siege to Fort Pitt. Seneca spread similar terror on the Allegheny; captured the smaller forts of Venango, Le Boeuf, and Presque Isle; and threatened to lay siege to Fort Niagara. In one month, the vaunted British military presence had almost disappeared west of the Appalachians. That left Fort Detroit.[4]

Regardless his role in the larger strategy, it made sense that Pontiac should concentrate his individual efforts against Fort Detroit. Not only was Fort Detroit in his immediate locale, it was also the hub of several smaller posts stretching from the Wabash north to Michilimackinac. Only Fort Niagara and Fort Pitt were of greater strategic value to the British west of the Appalachians.

Fort Detroit's commander was Major Henry Gladwin. Pontiac's presence at the Monongahela against Braddock might be open to question, but Gladwin had the scars to show that he had been there as a young lieutenant. Subsequently, he had acquitted himself well in Gage's newly raised light infantry regiment and served with Amherst's advance against Montreal. Amherst dispatched Gladwin westward in the summer of 1761 to assume command at Detroit and see to the permanent garrisoning of the other posts.

On the afternoon of Sunday, May 1, 1763, Pontiac and forty or fifty of his best warriors appeared at the east gate of Fort

Detroit. Gladwin treated them cordially, if somewhat cautiously, and the visitors undoubtedly used the occasion to reconnoiter the garrison under the guise of a ceremonial dance. Four days later, Pontiac held a war council with his own Ottawa and neighboring Huron and Potawatomi. The result was a plan whereby Pontiac would enter the fort in friendship with sixty warriors carrying hidden weapons. Other warriors would surround the fort and cut it off from any nearby reinforcements. When Pontiac gave the signal, bedlam would ensue and Gladwin's soldiers would be caught by surprise.

But there were flaws in Pontiac's plan. First, Fort Detroit held a much larger garrison than those smaller posts soon to be captured by similar ruses. Exact figures are unknown, but estimates suggest that it numbered about 120. Second, Major Gladwin was no dupe and after eight years in North American campaigns may well have suspected something amiss. Finally, there was the matter of an informant. At some point, Gladwin learned of Pontiac's planned deception.

Just who the informant was has long been one of history's little mysteries. Gladwin himself acknowledged the existence of such a source in his subsequent report to Amherst, saying, "I was luckily informed the night before" of the scheme. In one of the most romanticized versions, an Indian princess is in love with Gladwin and warns him. In other versions there are informers from among the locals—both Indians and French—as well as English traders who may have stumbled on the plan. In fact, Gladwin may have had more than one informant, and he would have been less than the officer he was if he had not suspected something on his own.

So, when Pontiac and his warriors entered the fort on May 7, 1763, they were the ones to be surprised. Gladwin's little garrison was turned out and fully armed on the parade ground. Gladwin still greeted his guest cordially, but Pontiac knew that his ruse would not work and retreated. The following day, he returned with only three Ottawa chiefs and attempted to dissuade Gladwin from believing in any ulterior motives. Tomorrow, Pontiac told

Gladwin, he would return with all of his warriors to show their respect. Pontiac may have felt that the guile worked, but Gladwin simply doubled his guard.

The next day, May 9, Pontiac led a flotilla of sixty-five canoes from the Ottawa town on the eastern side of the Detroit River. When he found the fort's gates closed and learned that Gladwin would permit only a few of his chiefs to accompany him inside, Pontiac at last knew that more than guile would be needed to carry Fort Detroit. Enraged, he returned to his village and divided his forces into raiding parties. If the fort could not be taken by ruse, he would destroy all the settlers around it and cut off any reinforcement. The first settlers to be killed were a Mrs. Turnbull and her two sons, who lived in a farmhouse about a mile from the fort.[5]

For the next two and a half months, a state of siege existed at Fort Detroit and its environs. This was certainly not the classic European siege involving trench warfare that had enveloped Fort William Henry and Louisbourg. Rather, it was a game of cat and mouse, in which Pontiac's warriors maintained a deadly encirclement and frequently pounced on parties that dared to sortie from the fort for reconnaissance or supplies.

Most dramatically, a force of almost 100 men bound for Fort Detroit was caught unaware of the hostilities and cut to pieces near the mouth of the Detroit River. But this had little strategic effect on Gladwin's well-stocked garrison. Through the summer, Pontiac failed to draw the noose tighter and refrained from a frontal assault. Such an assault might well have taken the fort, but at a loss of life to the Ottawa and their allies that would have been unacceptable in Indian warfare. For his part, Gladwin used two vessels in the Detroit River to bombard the Ottawa camp.

All this changed on the morning of July 28, when British sentries heard gunfire coming from the Huron camp downriver. At first, Gladwin suspected that Pontiac was finally beginning a major frontal assault. A dozen volunteers slipped out of the water gate of the fort and quietly paddled off in a bateau to reconnoiter the threat.

The river was cloaked in a thick fog, and the sounds of a large flotilla coming upriver could be heard long before anything was visible. Suddenly, one small vessel emerged from the fog, and then another and another. But these were not Pontiac's war canoes. Instead, they proved to be a fleet of bateaux, twenty-two in all, filled with red-coated British soldiers. Under the command of Captain James Dalyell, a reinforcement of 260 men—more than double the number of Major Gladwin's beleaguered garrison and including a contingent of rangers commanded by Robert Rogers himself—had reached Detroit. As the rising sun broke through the fog, the soldiers lining the walls of the fort raised a hearty cheer of welcome.

But increased numbers would not necessarily increase tactical acumen. Captain Dalyell was not cut from the same cloth as Major Gladwin, although his credentials were similar. Dalyell had entered service in North America in 1756 as a young lieutenant with the Royal Americans. Like Gladwin, Dalyell served with Gage's Eightieth Regiment of Light Infantry and by now should have been well versed in wilderness warfare.

But Dalyell had ambitions that had carried him into Amherst's inner circle as an aide-de-camp. Well aware of the general's increasing frustration with the news coming from the west, Dalyell saw the mission to relieve Fort Detroit and crush Pontiac as a way to promotion and greater glory. After arriving at Fort Detroit and resting his men for only two days, Dalyell urged an immediate attack on Pontiac's most recent encampment two miles to the north.

Nominally, of course, Major Gladwin remained in command of Fort Detroit. Gladwin strongly opposed such an offensive operation, preferring to let the dwindling summer wear out the patience of Pontiac's Huron and Potowatomi allies—some of whom were already showing signs of defection. But Dalyell flaunted his close ties to Amherst and insisted on an attack in force. At last Gladwin relented and approved a sortie in which Dalyell led most of the men he had brought with him. Major Rogers, who should have either known better or led the whole assault himself, went with the force.

In the early hours of July 31, 1763, Dalyell and his troops moved out of the east gate of Fort Detroit and marched two abreast up the river road toward Pontiac's camp. Two bateaux, each mounted with a swivel gun, shadowed the column from the river. A mile and a half from the fort, the long line formed into platoons with advance and rear guards, but the force was still spread out in a long column as it approached a narrow wooden bridge over Parent's Creek. Such a form of advance shows how little some British regulars—Dalyell most of all—had learned since Braddock's defeat. One can only wonder what Rogers thought of it.

Pontiac, of course, was hardly caught by surprise. Instead, he had assembled more than 400 warriors—mostly Ottawa and Chippewa—and divided them into two groups. Two hundred fifty warriors watched silently from the woods, and as Dalyell's column passed by, they positioned themselves across the road to cut off its retreat back to the fort. Pontiac and 160 others hid themselves in a semicircle on the far side of the bridge. Pontiac's strategy was clearly to surround and annihilate as many of his enemies as possible.

Inexplicably, Dalyell's advance guard "tramped noisily over the planks" and then were hit by a hail of musket fire from the concealed warriors. As the advance guard crumpled, the main body, consisting of several hundred troops, was exposed to a withering fire from three sides, reminiscent of that day on the Monongahela. What on earth had Dalyell been thinking? Now, sounds from the rear of his force indicated that he was surrounded. At the first sound of the ambush, only Major Rogers had the presence of mind to scurry some of his rangers to the protection of a nearby farmhouse and effectively turn it into a blockhouse from which to anchor one flank.

As the moonlight slowly turned to dawn, a furious battle continued. Dalyell, though fighting bravely, seemed uncertain whether to press forward or fight a retreat back to Fort Detroit. Finally, he chose the latter and personally led a charge to drive through the warriors blocking the escape route back to the fort. The effort cost Dalyell his life, but saved what remained of his troops.

With Rogers covering the retreat from his improvised block-house and then falling back himself, the tattered force stumbled back to Fort Detroit. Dalyell and nineteen men were dead and at least forty others wounded. Pontiac's losses are unknown but are said to have been seven killed and a dozen wounded. Several of the dead British soldiers fell into Parent's Creek and turned its waters red with their blood. Henceforth, the stream has been known as Bloody Run.

Despite Dalyell's tactical blunder, the battle of Bloody Run was hardly a great strategic victory for Pontiac. First, he had failed to annihilate the British force, and as a result Gladwin's garrison stood greatly reinforced. A week later, the schooner *Huron* arrived with sixty more men and eighty barrels of provisions. Clearly, Fort Detroit still stood firm. Absent a frontal attack, which Pontiac had refused to make even when he held overwhelming superiority, there was no alternative but to continue his loosely enveloping siege.[6]

Meanwhile, a veteran of Forbes's Road, Colonel Henry Bouquet, was marching to the relief of Fort Pitt. He had been in Philadelphia arranging for supplies when he first received reports of trouble not only at the forks but also as far east as Fort Bedford at Raystown. Bouquet was certainly well acquainted with the logistics of the supply line between Philadelphia and Fort Pitt, as well as the internal politics of the Pennsylvania legislature, which now once again spent crucial weeks debating the merits of raising provincial companies to aid in the campaign.

Finally, it was with two decidedly understrength regiments of Highlanders, a battalion of Royal Americans, and a company of rangers—some 460 men—that Bouquet marched west from Carlisle on July 18, 1763, without Pennsylvania troops. "Though I feel myself utterly abandoned by the very people I am ordered to protect," wrote Bouquet, "I shall do my utmost to save them from destruction."[7]

And destruction was what Amherst had told Bouquet to visit on his foes. The commander in chief was already frustrated by his lengthy stay in North America, and his normally calm demeanor rapidly turned to impatience and then rage as reports of the fall of one post after another streamed into his headquarters in New York. Take no prisoners, Amherst advised Bouquet, only to learn from Bouquet in return that Le Boeuf, Venango, and Presque Isle had also fallen.

Then, Amherst made an even more startling suggestion to his field commander. "Could it not be contrived," the general asked, "to send the small pox among the disaffected tribes of Indians? We must on this occasion use every strategem in our power to reduce them." Bouquet replied that he would do so, and there is evidence that blankets infected with smallpox were distributed to Indians in parleys at Fort Pitt even before Bouquet's relief force arrived.[8]

What level of enthusiasm Bouquet, always the good soldier, brought to this tactic is open to question, in part because Bouquet had issued orders supporting the Treaty of Easton and prohibiting certain settlements even before the Proclamation of 1763. Fortunately for Amherst, Bouquet's strict sense of duty in this regard also made him immune to certain temptations of the frontier. Bouquet flatly rejected an offer of 25,000 acres from the Ohio Company—an outright bribe—if he would overlook the establishment of certain settlements on the Ohio by families from Virginia and Maryland.[9]

Leaving some of his small force to strengthen the garrison at Fort Bedford, Bouquet moved westward to Fort Ligonier. Here, there was no recent news from beleaguered Fort Pitt, only forty miles away. In fact, the last week of July had seen a ferocious attack against Fort Pitt that had tried to take it by storm. Not only was this frontal attack in contrast to the tactics Pontiac was applying at Detroit, but the siege of Fort Pitt was also a much tighter noose. By August 1, the Delaware and Shawnee attackers had dug rifle pits along the river and were sniping at soldiers on the parapets.

Having no firm intelligence of this but nonetheless surmising Pitt's plight, Bouquet elected to leave his heavy wagons and ammunition train at Fort Ligonier and hurry toward the forks with 300 men and 340 horses loaded with flour. At first glance, it seemed a move similar to Braddock's division of command and hasty rush forward eight years before. Fort Pitt's attackers heard of Bouquet's advance and broke off their siege to lay an ambush for his column about twenty miles west of Fort Ligonier on the hills above a small stream called Bushy Run.

On the afternoon of August 4, 1763, Delaware and Shawnee warriors attacked the advance units of Bouquet's command. Sensing far more than a skirmish, the colonel deployed his troops into a circle and ordered them to use the heavy flour sacks for cover. All spent a generally unpleasant night. The following morning, the attack was renewed with increased vigor, but far from feeling surrounded, the seasoned Bouquet had a trick or two up his sleeve. He ordered two companies of light infantry to fall back from their position at the head of his circled position. Thinking it was a retreat, the Delaware and Shawnee followed them.

The Indian attackers were quickly counterattacked as planned by musket volleys on both sides and a deadly bayonet charge. Those warriors who escaped the circle retreated in confusion, leaving among their dead the Delaware chief Tesekumme, who had been their chief negotiator. Five days later, Bouquet and his men marched into Fort Pitt badly battered but unopposed. The power of the Ohio Indians was far from destroyed, but Fort Pitt would remain in British hands for the coming winter.[10]

As it turned out, Fort Detroit, too, would remain in British hands. "The enemy are masters of the country, the season far advanced," Gladwin reported to Amherst early in October, "not a stick of fire wood in the garrison, and but little provisions." But Pontiac was also short of supplies. As the autumn days grew shorter, more and more of his warriors slipped away to hunt and prepare for the winter. Finally, the Ottawa chief gave up the siege of Detroit and went south himself.[11]

So it was that as the year 1763 ended, the British found themselves with a tenuous grip on the Great Lakes country, and the pivotal posts of Niagara, Pitt, and Detroit. By the following spring, Jeffery Amherst was finally back in England, but Thomas Gage would carry out his former superior's charge to strike firmly against all Indians who had dared to defy him. Sir William Johnson convened a peace conference at Fort Niagara in July 1764; and then the old bateau man John Bradstreet, now a major general, led a force of 1,400 west to reinforce Detroit.

Meanwhile, the stalwart Henry Bouquet embarked with 1,500 men—including the long-awaited Pennsylvania provincials—on a similar mission west from Fort Pitt. Leaving on October 3, 1764, incredibly late in the season, Bouquet circled through what is now southern Ohio and momentarily established peace with the Ohio Indians under threat of burning their villages. "We have visibly brought upon us this Indian war by being too saving of a few presents to the savages which properly distributed would certainly have prevented it," Bouquet reported to Gage. But nothing could prevent the rush of westward expansion into the vast lands called an "Indian reserve." This unresolved matter, as well as Bradstreet's and Bouquet's expeditions of 1764, were far more a prelude to the next era than a conclusion to the previous one.[12]

Whether one characterizes the Indian attacks of 1763 as a grand conspiracy or—more nobly—part of a Native American war for independence, armed conflicts between Indians and European settlers would continue in North America until a place called Wounded Knee, a century and a quarter later. France, Great Britain, and Spain might fight over the continent and decree its ownership among themselves, but in hindsight there can be little doubt who were the real losers in the French and Indian War.

But if Native Americans were disgruntled in the wake of the war, colonials in the thirteen colonies—seventeen colonies if one counted Nova Scotia and the new additions of East and West Florida and Quebec—soon found that they had their own reasons to be disgruntled. The presence of British regulars actually increased after the war, and their quartering in private homes when there was no impending threat became a source of contention. Taxes were levied—without representation—to pay for the costs of William Pitt's global war. Most grating of all, being forced to pay for his majesty's conquests, the colonials were now denied a share in the rewards by the prohibition of settlement west of the Appalachians. In some respects, it was almost worse than if they were still encircled by the French.

In the end, there was one more casualty of the war. In the spring of 1758, as Jeffery Amherst hastened to Louisbourg to do William Pitt's bidding, Pitt had assured him that he could return to England as soon as practical. But there had always been some exigency to prevent it. Finally, in the fall of 1763—five and a half years after Amherst had left his wife and home in England—the general received permission to return "as soon as the Indian uprising was disposed of." Such a disposition would prove subjective, of course, but in November 1763, Major General Jeffery Amherst made the determination that it was over and embarked for England, turning over his position as commander in chief of British forces in North America to Thomas Gage, now a major general.

Amherst had definitely overstayed his welcome in North America, but what he found in England was both personally and professionally devastating. His beloved wife, Jane, had gone insane during his long absence and would die within a few months. His country estate at Riverhead was a shambles. Halfheartedly, he soon razed it to build a house nearby called Montreal, but this gave him little solace. Professionally, there were no accolades, no parades, no parliamentary testimonials such as had greeted the coffin of his young protégé, James Wolfe. "A new king was on the throne, a new ministry in power, and Amherst's brilliant victories of 1759 and 1760 belonged to a war now ended. Instead, the Indian revolt in the West was in the news, and Amherst was pointed out as the general who had failed to subdue it."[13]

Writing to a close friend, Amherst confessed, "I may tell you for your own information only, that I have no thought of returning to America." Fifteen years later, George III summoned

Amherst to Saint James Palace. The king's North American colonies were in revolt. "Gentleman Johnny" Burgoyne had just lost one British army at Saratoga and Lord William Howe, who had once led the way for Wolfe onto the Plains of Abraham, was bottled up in Philadelphia with another. Once again, Amherst's king appointed him commander in chief of British forces in North America. This time, Jeffery Amherst politely but firmly declined.[14]

20

PRELUDE TO REVOLUTION

The triumphs of one global war—no matter how satisfying to some—had nonetheless sown seeds of discontent that would soon lead to a bitter quarrel within the British Empire. While there were to be many reasons for it, high on the list was a lack of any forceful, informed leadership in the highest echelons of George III's government. Barely was the ink dry on the Treaty of Paris when Lord Bute resigned, in part because he quickly realized that it was one thing to be the power behind the throne and quite another to suffer the criticism that came with being its public face. George Grenville, another of William Pitt's brothers-in-law, succeeded Bute in April 1763. But Grenville was hardly Pitt's ally, and events during Grenville's short ministry only served to alienate the two men further.

Great Britain was short of cash. There was no denying that. Expenses had burgeoned during Pitt's wartime tenure, and with the new demands of administering a global empire, they were unlikely to shrink. In order to raise revenues to meet these outlays, Grenville championed the passage through Parliament of what has traditionally been called the Sugar Act.

Technically, this legislation was the American Duties Act of 1764, and it not only placed tariffs on sugar, coffee, wine, and many other imports into Britain's North American colonies, but also imposed increased enforcement measures to end smuggling and collect all taxes due. Such taxes were certainly not new—a tax of sixpence per gallon had been imposed on foreign molasses as early as 1733—but the announced resolve to enforce them firmly was an entirely different matter. "The publication of orders for the strict execution of the customs laws," observed the governor of Massachusetts, "caused a greater alarm in this country than the taking of Fort William Henry in 1757."[1]

Then, Grenville dropped the other shoe. The Stamp Act of 1765 was another measure designed to raise revenue. Essentially, it was a tax on paper. Any paper, including newspapers, licenses, playing cards, legal documents such as wills and powers of attorney, and a host of other printed matter, was to have a revenue stamp affixed to it certifying that the appropriate tax had been paid. Passed by Parliament in April 1765 with assurances that the revenue derived would be used to defray the expenses of defending and administering the colonies, the new measure was supposed to go into effect November 1. Unlike the Sugar Act, which flowed from Parliament's uncontested power to regulate commerce, the Stamp Act was a direct tax on individuals—without representation—and colonials howled in indignation.

Among the colonial firebrands was a young lawyer named Patrick Henry, who went so far as to suggest to the Virginia House of Burgesses that Parliament "had no legal authority to tax the colonies at all." Warned that his words bordered on treason, Henry made his famous reply that if "this be treason," the delegates should make the most of it.

Assemblymen in Massachusetts were only slightly more restrained and called for an intercolonial convention, the like of which had not been seen since the Albany Congress eleven years before. Twenty-seven delegates from nine colonies met in New York City in October for the Stamp Act Congress. But the pronouncements of Patrick Henry and this assemblage paled compared with the outpouring of protest from the broad spectrum of the colonial population. They not only refused to buy stamps but also boycotted British imports. This boycott did nothing, of course, to stimulate a prosperous postwar trade.

"What used to be the pride of the Americans?" Benjamin Franklin was asked when called before the House of Commons to testify on the Stamp Act and the resulting boycott. "To indulge in the fashions and manufactures of Great Britain," the best-known colonial in London replied. "And what is now their pride?" he was asked. "To wear their old clothes over again, till they can make new ones."[2]

By the end of the year, a lawyer in Braintree, Massachusetts, named John Adams confided to his diary that "the great men [in London] are exceedingly irritated at the tumults in America, and are determined to enforce the act. This irritable race, however, will have good luck to enforce it. They will find it a more obstinate war, than the conquest of Canada and Louisiana."[3]

At least some members of Parliament agreed. "I rejoice that America has resisted," proclaimed a battered and infirm William Pitt in the House of Commons. "In a good cause, on a sound bottom," he continued, "the force of this country can crush America to atoms. . . . But on this ground, on the Stamp Act, when so many here will think it a crying injustice, I am one who will lift up my hands against it."[4] Repeal the Stamp Act, Pitt urged, and on March 4, 1766, Parliament did just that.

But important lessons had been learned by the colonials in North America, who had seen the power of concerted action. Equally important lessons had not been learned by Parliament. On the same day that it voted the repeal of the Stamp Act, Parliament overwhelmingly passed the Declaratory Act. Having just recognized certain limits to empire and yielded to colonial pressure to repeal the Stamp Act, Parliament nonetheless sought to reassert itself. It declared Parliament's full right and authority to make any and all laws and statutes governing those same colonies. The law would have little impact because colonial assemblies would soon choose to ignore other laws and taxes, but the mindset of those passing the Declaratory Act would prove an increasing impediment to mutually beneficial relations within the empire.

Finally, there was one other pressing issue of colonial concern.

Parliament had long routinely passed an annual multifaceted bill governing the administration of Great Britain's army. Called the Mutiny Act, it nonetheless covered all topics from recruitment to discharge. It also provided rules by which troops might be housed—quartered—in private dwellings. (The requirement to reauthorize the Mutiny Act annually stemmed from England's long-held bias against maintaining a standing army.)

Generally, the Mutiny Act applied only to the British Isles, and commanders in chief in North America from Braddock onward had been compelled to appeal to individual colonial legislatures to pass their own local versions. This they did with little complaint during the war. After the war, there were still reasons for the British government to maintain a standing army in North America—Pontiac had proved that. But when Major General Thomas Gage wanted to withdraw regular troops from far-flung frontier posts and garrison them in major cities in response to civilian unrest over the various tax acts, that was an entirely different matter. After the mayor of New York refused to provide firewood for regulars quartered in that city, Gage appealed to Parliament to extend the Mutiny Act to North America.

Interestingly enough, the Quartering Act of 1765, that was passed as a result, applied only to vacant private buildings and arguably exempted occupied private dwellings. But the act also authorized the impressment of wagons at the going rate; required men billeted in public houses to pay only for food, not lodging; and allowed troops to use river ferries at half fare. It didn't take long for colonials to argue—particularly in New York, where Gage had his headquarters—that such perquisites were really hidden taxes on teamsters, innkeepers, and ferrymen. New York took the lead in nullifying the Quartering Act and refused to obey it. Many in Great Britain were aghast. If a colony could nullify one act of Parliament, what was to stop it from nullifying others? Indeed, how did such action square with the very term "colony" and the supposed sovereignty of the British crown?[5]

George Grenville survived two uneasy years at the helm amid

this sort of turmoil and then was succeeded for a year by the marquis of Rockingham. Finally, recognizing that his ship of state was largely rudderless, George III swallowed his pride in the summer of 1766 and asked William Pitt to form a government once more. The choice of Pitt was immensely popular in North America, where his opposition to the Stamp Act had made him a hero. At home, the British people waited for the "great commoner" to work his old magic.

But Pitt was well past his prime, worn out by illness and able to rise to only occasional flashes of his previous brilliance. Then, in order to support the king's initiatives in the House of Lords, Pitt accepted a peerage and became the earl of Chatham. Many judged that this was not his finest hour. The British middle class, who had long supported him, felt betrayed. Pitt himself descended into a dark depression that left him absent from the cabinet more than he was present.

Since George III's government was still faced with a void in leadership, it is small wonder that the solutions subsequently applied to North America were ill-advisedly conceived and brashly executed. Urging them on was Pitt's chancellor of the exchequer, Charles Townshend, the nephew of the duke of Newcastle who had embarked for Quebec as one of Wolfe's brigadiers in 1759 to add a little military luster to his political record.

Townshend used Pitt's infirmity to rush to the forefront as his possible successor, but what came to be called the Townshend Acts—new taxes on glass, lead, paints, paper, and tea imported into the colonies along with additional enforcement measures—showed that he had learned nothing from Pitt. "Champagne Charlie," as Townshend was called because of his high living and snobbery, viewed the Americans as "ungrateful brats" to be chastised. This attitude was a far cry from Pitt's assertion during the crisis over the Stamp Act that "the Americans are the sons, not the bastards of England." By the end of 1767, Townshend was unexpectedly dead, but the animosity that his acts caused would live on.[6]

Pitt's ill-fated second ministry was followed in 1768 by that of

Lord Grafton, who lasted little more than a year; and then by that of Lord North, who proved as intensely loyal to George III as he was contemptuous of colonials in North America. North would remain as prime minister until the seeds of discord in the colonies had sprouted into full-grown trees of revolution. Taxes on tea, customs restrictions, and the quartering of troops would all continue to rankle, but it was the Quebec Act of 1774 that convinced colonials that their fear of encirclement by the French had been replaced by fear of an even more oppressive force.

Enacted as one of the Intolerable Acts, the Quebec Act expanded the boundaries of the colony of Quebec to include all lands north of the Ohio River. These were the very lands to which young George Washington had advocated Virginia's claims on the cold December day in 1753, and which the subsequent Proclamation of 1763 had supposedly set aside as an Indian reserve. The proclamation had never been rescinded, but even Washington observed that the line "seems to have been considered by government as a temporary expedient . . . and no further regard has been paid to it."

In other words, with a wink and a nod, Virginia, Maryland, Pennsylvania, and New York had quietly continued their claims to these lands. But while the Proclamation of 1763 ostensibly prohibited westward expansion even as it overlooked continued settlement schemes, "the Quebec Act destroyed such schemes permanently" and made these lands part of the colony of Quebec.[7]

The Quebec Act also mandated a governor and council appointed by the crown for Quebec, denied an elected assembly, and permitted the dominance of the Roman Catholic church. The last of these provisions was meant to appease French Canadians, but was seen by many colonial Protestants as a popish plot. The other Intolerable Acts included a far more onerous Quartering Act and the full commercial closure of the port of Boston until payment was made for a certain quantity of tea that some of its citizens had dumped into Boston Harbor one night. All this came to a head on April 18, in 1775, when two lanterns were hung in the steeple of Boston's old North Church and their beams sent messengers riding toward Lexington and Concord.

By then, where were those who had played leading roles in what William Pitt had once hoped would be Great Britain's great war for lasting empire? The former governor of Massachusetts, William Shirley, had retired as governor of the Bahamas and died in Roxbury, Massachusetts, in 1771. He was spared the coming violence that seemed so counter to his congenial relations with the Massachusetts Assembly, but before his death he had professed bewilderment with the many new faces. When asked about certain young troublemakers in Massachusetts, Shirley burst out: "Mr. Cushing I knew, and Mr. Hancock I knew; but where the devil this brace of Adamses came from, I know not."[8]

Lord Howe's youngest brother, William, arrived back in Massachusetts in 1772 as a major general. He was sent there to reinforce Thomas Gage's efforts to suppress the unruly populace of Boston. Howe found the task embarrassing if not actually distasteful, because these same citizens had recently commissioned a monument to his late brother for Westminster Abbey. When Gage became the scapegoat for the British blunder at Bunker Hill, William Howe succeeded him as commander in chief of British forces in North America.

William Howe's opposite number, of course, was George Washington, who, despite a record of defeat and some surliness if not outright insubordination, had emerged from the French and Indian War as Virginia's favorite military son. Howe would spend the next two and a half years sparring with Washington around New York and Philadelphia, but after a victory at Brandywine, he failed to stamp out the flame left to burn at Valley Forge during the winter of 1777–1778. Howe was recalled to England thereafter, and history would be left to ponder whether his inactivity was born of mere incompetence or a measure of colonial affection shared with his late brother.

The object of the first Lord Howe's admiration, Robert

Rogers of the rangers, did not live up to his legend. In 1766, Rogers received the command of Fort Michilimackinac, from where he hoped to profit from the fur trade and search for the Northwest Passage. Rogers sent out two exploring parties, but neither got very far. The real problem, however, was that Rogers was still not getting along well with regular officers. Charges of his favoring certain Montreal merchants for commercial gain accumulated until Rogers was accused of treason in planning to establish his own country west of the Great Lakes. Ignominiously, Rogers spent a winter in irons and then was hauled east. He was eventually acquitted but never returned west

By 1775, Rogers was in considerable debt and was more concerned with seeking land grants in compensation for his earlier services than with aiding colonial efforts. Neither the colonials nor the British were quite sure which side he was on. Quite possibly, Rogers wasn't sure himself. Finally, he agreed to recruit British loyalists for the Queen's American Rangers. His service was sporadic and gained him little but two heartbreaks. His wife, Elizabeth, who had stood beside him through poverty and the humiliation of the winter at Fort Michilimackinac, finally divorced him on charges of desertion and cruelty. Then, the New Hampshire legislature—wanting no Tories on its soil—banned him from New Hampshire. Major Robert Rogers, the dashing hero of one era, died broke and alone in another era, on May 18, 1795, in England. By one account, his shouts of names and places long since past punctuated his final fits of drunkenness.[9]

The Ottawa chief Pontiac fared little better. Pontiac avoided the punitive campaigns of Bouquet and Bradstreet in 1764 by hiding out in the Illinois country. When Bradstreet arrived in Detroit, he considered demanding that the Ottawa hand Pontiac over as a condition of peace, but then thought better of it. For his part, Pontiac seems never to have reconciled himself to Great Britain's new dominance, and rumors circulated for several years that he was plotting more warfare. In the summer of 1766, Pontiac traveled east with a delegation representing many Indian nations and met with Sir William Johnson at Oswego. In the negotiations that followed, Johnson seems to have overplayed Pontiac's importance among the assembled delegates. Their jealousy was further fueled by rumors that Pontiac was to have a pension for his role in pledging "peace and friendship with Great Britain."

But the British view of his importance was not what Pontiac found when he returned to the west. In some ways, trouble seems to have dogged him as it did Robert Rogers. No one was sure whether Pontiac was plotting with Spain and France or was truly a new ally of Great Britain. Within two years, Pontiac was in fact reduced to wandering the Illinois country with little influence, but Peoria Indians living near Cahokia were taking no chances. On April 20, 1769, as Pontiac left a trading post in Cahokia, a Peoria murdered him. The very fact that this act caused no retaliation suggests that this warrior, too, had outlived his time.[10]

Governor General Vaudreuil was sent to France to account for his role in the fall of New France. It soon became clear that his dissension with Montcalm had extended beyond military matters. Montcalm had gathered considerable evidence of the governor's involvement in a ring of corruption that had bled what little resources the colony possessed into private pockets. Montcalm appears to have entrusted much of this evidence to the care of Father Roubaud at the mission of Saint Francis. Roubaud was absent on the day that Rogers' Rangers set the village ablaze, and the Vaudreuil papers were evidently destroyed. Without hard evidence against him, Vaudreuil was eventually exonerated, but several of his chief conspirators were not so lucky and ended up in the Bastille. Vaudreuil never returned to Canada; he spent a peaceful retirement in Paris until his death in 1778.[11]

Montcalm's trusted aide-de-camp Louis-Antoine de Bougainville rose above the muddle of New France and, interestingly, left France's army to join its navy. Bougainville conceived a plan to settle displaced French Canadians on the Falkland Islands, then not claimed by any European power. After Spain protested, however, Bougainville sailed from France first to surrender the Falklands to

Spain and then to continue on an epic three-year voyage around the world in 1766–1769. He became the first French naval officer to do so, but his lasting fame may come from the multiblossomed plants that he discovered, *Bougainvillea*. Later, when French and British fleets clashed off Yorktown in 1781, Bougainville was there in command of a French ship of the line.

And what of William Pitt? Though a shadow of his former self, the "great commoner," now ensconced in the House of Lords, continued to lift his weakened voice in support of America. He urged the removal of troops from Boston, abhorred the policies of Lord North, and confessed that were he but ten years younger, he would "spend the remainder of my days in America, which has already given the most brilliant proofs of its independent spirit." On May 11, 1778, William Pitt breathed his last. George III professed surprise at the vote for his public funeral and continued to carry on the war that Pitt had opposed.[12]

All this raises some interesting "what ifs." History is filled with such questions, and that is particularly true of the period which decided the fate of a continent and set up the American struggle for independence. What if Lord Howe had not been killed at Fort Carillon? Might his survival have led to a far gentler and more understanding administration of Massachusetts than the one Thomas Gage dictated? And would Lord Howe have been a more competent military leader against Washington than his brother, William, if it had come to that?

What if the marquise de Pompadour had been more interested in North America and a global vision for France than she was in being flattered by Maria Theresa? Indeed, what if *anyone* in France had been more interested in a global vision and had nurtured it with naval resources? What if there had been a counselor for Louis XV of the stature of his ancestors' Richelieu or Mazarin? By the time Choiseul consolidated power, it was too late, and his sole interest was in saving France—with or without Canada.

What if Spain had entered the war much earlier and thrown its naval weight against Great Britain in the critical early years, uniting the Bourbon thrones despite the dictate of the Treaty of Utrecht? Might a combined Spanish and French fleet have been able to control the English Channel long enough to support an invasion of England? True, this had not been accomplished since 1066, and a Spanish armada had failed once before; but even Pitt worried about the possibility.

What if Pontiac had been possessed of the full command and control capabilities that Parkman imputed to him? Might a Native American alliance from the Iroquois across the Great Lakes to beyond the Mississippi have stopped British expansion cold and forced on the thirteen colonies the same measure of dependence on Great Britain for protection from their encirclement that they had required for a century against New France?

What if a few key military matters had been reversed? Might the British have prevailed in North America much sooner if Abercromby had brought up his artillery instead of rushing the barricades at Fort Carillon? And might the British not have prevailed at all if Wolfe's wave of a hat in the surf of Gabarus Bay had indeed signaled retreat or if Montcalm's patience had given Bougainville an hour to close a trap on the Plains of Abraham? Lacking that, what if the ships arriving off Quebec in the spring of 1760 had been flying the fleur-de-lis? And on it goes.

But these "what ifs" belong to conjecture. What is fact is the far-ranging historical significance of the French and Indian War and the greater Seven Years' War of which it was a part.

The French and Indian War created the British Empire. Admiral Mahan was indeed correct. After the war, the kingdom of Great Britain became the British Empire. Neither the repudiation in the Treaty of Paris of many of Pitt's territorial conquests nor the subsequent fiasco of Britain's colonial administration in North America could change that. The empire that Pitt won from Quebec to India would dominate world affairs for the next two centuries.

The French and Indian War decided the fate of the North

American continent among Great Britain, France, Spain, and Native Americans. The war decisively expelled France from North America, although descendants of the French and French culture flourish to this day in the province of Quebec, in part because of the Quebec Act of 1774 that so outraged English colonists. By giving Quebec's residents—both British settlers who had arrived in the aftermath of the war and French Canadians who had stayed—room to expand and freedom of religion, Great Britain ensured their loyalty. Quebec remained firmly in the British Empire, not only against the political and military entreaties of the thirteen colonies during the American Revolution, but also later during the War of 1812.

Spain's late allegiance to France in the French and Indian War cost it Florida. Although Spain would briefly recover Florida after the American Revolution, its quest for a North American empire reached a zenith in 1763. Had Spain belatedly prevailed against Portugal and acquired Brazil, it might have emerged from the war invigorated rather than commercially and economically drained. By 1810, Spain's colony of Mexico would follow the example of the thirteen colonies and begin its own campaign for independence from a European throne.

And what of the Native Americans? They, too, lost what had been their continent to Great Britain. Another century and a quarter of warfare would ensue between the conquerors and the vanquished from the swamps of Florida to the plains of Saskatchewan. But after the French and Indian War, never again would Native Americans in North America present so concerted and influential a force as the Iroquois Confederacy had done before 1763. Had Pontiac truly lived up to his legend and been able to combine the energies of the upper Mississippi Indians with the Six Nations and their vassals, the westward march of Europeans might have at least been delayed.

Finally, the French and Indian War—and Great Britain's need to pay for it—precipitated the issues that led to the American Revolution: taxation without representation; trade and customs regulations; quartering troops in private homes; and restrictions on westward expansion. The war also proved that however disjointed and premature the discussions of colonial union had been at Albany in 1754, it had—western land rivalries aside—slowly fostered some measure of intercolonial cooperation. In doing so, it also trained a cadre of young provincial officers from George Washington to Francis Marion, who, having learned to fight for one sovereign, would continue to fight for the sovereignty of all. Great Britain indeed won a continent, but in doing so, it lit the spark of revolution.

If there is an epilogue to France's loss of North America in the French and Indian War, perhaps it came on a spring day in 1778 at a small place just northwest of Philadelphia called Valley Forge. On that day, George Washington's tiny colonial army was drawn up on the parade ground under the watchful eye of an expatriate Prussian drillmaster. Three cheers were given. The occasion was the announcement that the fledgling union of the thirteen colonies had an international ally. Thanks in no small measure to the entreaties of Benjamin Franklin, France had formally recognized their sovereignty and pledged to support them in a war against Great Britain. Ironically, three and a half years later, it was French naval power that sealed the fate of Cornwallis's army at Yorktown by defeating British naval forces off the Virginia capes. In the end, the French navy that had lost a continent helped to win one for the young American nation.

NOTES

1 • The Bells of Aix-la-Chapelle

1. Francis Parkman, *France and England in North America*, Vol. 1., *La Salle and the Discovery of the Great West* (New York: The Library of America/Viking, 1983), p. 927. (Reprint.)
2. Samuel Eliot Morison, *The Oxford History of the American People* (New York: Oxford University Press, 1965), p. 122. (Note: All contemporary quotations have been corrected for spelling and capitalization to facilitate reading.)
3. Ibid., p. 137.
4. Ibid., p. 139.
5. Ibid., pp. 155–156.
6. Charlton W. Tebeau, *A History of Florida* (Coral Gables, Fla.: University of Miami Press, 1971), pp. 69–70.
7. "Our trade will improve": Walter L. Dorn, *Competition for Empire, 1740–1763* (New York: Harper, 1940), p. 9. "Peace without victory": T. Walter Wallbank et al., *Civilization: Past and Present* (Glenview, Ill.: Scott, Foresman, 1967), p. 435.

2 • Beautiful Ohio

1. Francis Parkman, *Montcalm and Wolfe: The French and Indian War* (New York: Da Capo, 1995), pp. 12–13. (Reprint.)
2. Fred Anderson, *Crucible of War: The Seven Years' War and the Fate of Empire in British North America, 1754–1766* (New York: Knopf, 2000), pp. 14, 21; Wilbur R. Jacobs, *Wilderness Politics and Indian Gifts: The Northern Colonial Frontier, 1748–1763* (Lincoln: University of Nebraska Press, 1966), p. 27.

3. "Pierre-Joseph Céloron de Blainville," *Dictionary of Canadian Biography*, Vol. 3. *1741 to 1770* (Toronto: University of Toronto Press, 1974), pp. 99–100; A. A. Lambing, ed., "Céloron's Journal," *Ohio Archaeological and Historical Quarterly*, Vol. 29, No. 3 (July 1920), pp. 336–340, hereafter "Céloron Journal." "In certain places": A. A. Lambing, ed., "Account of the Voyage on the Beautiful River Made in 1749, under the Direction of Monsieur De Céloron, by Father Bonnecamps," *Ohio Archaeological and Historical Quarterly*, Vol. 29, No. 3 (July 1920), pp. 401–402, hereafter "Bonnecamps Journal."

4. "Céloron Journal," p. 341.

5. Ibid., pp. 348–351. (There is considerable local debate over the exact location of the village of Attiqué.)

6. Howard H. Peckham, *The Colonial Wars: 1689–1762* (Chicago, Ill.: University of Chicago Press, 1964), pp. 124–125.

7. Peckham, *Colonial Wars*, p. 125. "Maintained therein": "Céloron Journal," pp. 341, 371. "So little known": "Bonnecamps Journal," p. 409.

8. Parkman, *Montcalm and Wolfe*, p. 30.

9. Thomas D. Clark, *Frontier America: The Story of the Westward Movement*, 2nd ed. (New York: Scribner, 1969), pp. 24–34, 39.

10. Peckham, *Colonial Wars*, pp. 128–129. "We don't know": *Documents Relating to the Colonial History of the State of New York* (Albany: Weed, Parsons, 1855), Vol. 6, p. 813, hereafter *New York Colonial Documents* (Iroquois Red Head to Warraghüyagee—William Johnson—September 10, 1753).

11. Peckham, *Colonial Wars*, p. 130.

12. George Washington, *The Journal of George Washington: An Account of His First Official Mission, Made As Emissary from the Governor of Virginia to the Commandant of the French Forces on the Ohio, October 1753–January 1754* (Williamsburg, Va.: Colonial Williamsburg, 1959), p. 4. (Facsimile reprint of 1754 Hunter edition.)

13. Ibid., p. 4.

14. Ibid., p. 13.

15. Ibid., p. 16.

16. Ibid., pp. 25–26 (Dinwiddie to Commandant of French Forces on the Ohio, October 31, 1753).

17. Ibid., pp. 27–28 (Legardeur to Dinwiddie, December 15, 1753).

18. Ibid., p. 2.

3 • Albany, 1754

1. Adolph B. Benson, *The America of 1750: Peter Kalm's Travels in North America* (New Haven, Conn.: Yale University Press, 1937), pp. 340–344 (from a description of June 21, 1749). See also New York State Museum Web site for Albany at www.nysm.nysed.gov/albany; a map of early Albany in David R. Starbuck, *The Great Warpath: British Military Sites from Albany to Crown Point* (Hanover, N.H.: University Press of New England, 1999), p. 8; and the population estimate for 1754 in Catherine Drinker Bowen, *The Most Dangerous Man in America* (Boston, Mass.: Little, Brown, 1974), p. 99.

2. Charles Henry Lincoln, ed., *Correspondence of William Shirley* (New York: Macmillan, 1912), Vol. 2, pp. 13–14 (Lords of Trade to Shirley, September 18, 1753); hereafter *Shirley Papers*.

3. Wilbur R. Jacobs, *Wilderness Politics and Indian Gifts: The Northern Colonial Frontier, 1748–1763* (Lincoln: University of Nebraska Press, 1966), pp. 5, 11.

4. *Shirley Papers*, Vol. 2, pp. 12–13 (Earl of Holderness to Shirley, August 28, 1753).

5. "In case any": Ibid., pp. 14–15 (Shirley to Sharpe, November 26, 1753). "Than a well": p. 30 (Shirley to Lords Commissioners, January 1754).

6. Ibid., pp. 43–44 (speech to general court of Massachusetts, April 2, 1754).

7. Matthew C. Ward, *Breaking the Backcountry: The Seven Years' War in Virginia and Pennsylvania, 1754–1765* (Pittsburgh, Pa.: University of Pittsburgh Press, 2003), pp. 23–24.

8. *Shirley Papers*, Vol. 2, p. 46 (speech to general court of Massachusetts, April 2, 1754).

9. Samuel Eliot Morison, *Oxford History of the American People* (New York: Oxford University Press, 1965), pp. 69, 76, 108–109. Howard H. Peckham, *The Colonial Wars: 1689–1762* (Chicago, Ill.: University of Chicago Press, 1964), pp. 32, 70, 106, 116. "Were there a general": Leonard W. Larabee, *The Papers of Benjamin Franklin*, Vol. 4, *July 1, 1750, through June 30, 1753* (New Haven, Conn.: Yale University Press, 1961), p. 119 (Franklin to James Parker, March 20, 1751).

10. John A. Garraty, *The American Nation: A History of the United States* (New York: Harper and Row, 1966), pp. 89–96.

11. *Pennsylvania Gazette*, May 9, 1754.

12. Armand Francis Lucier, *French and Indian War Notices Abstracted from Colonial Newspapers*, Vol. 1, *1754–1755* (Bowie, Md.: Heritage, 1999), pp. 53–54.

13. *Ibid.*, p. 68.

14. Lawrence Henry Gipson. *The British Empire before the American Revolution*, Vol. 5, *The Great Lakes Frontier, Canada, the West Indies, India, 1748–1754* (New York: Knopf, 1942), pp. 113–114.

15. Ibid., *The Great Lakes Frontier*, pp. 119–122; Julian P. Boyd, *The Susquehannah Company Papers*, Vol. 1, *1750–1755* (Ithaca, N.Y.: Cornell University Press, 1962), pp. 101–103.

16. Larabee, *Papers of Benjamin Franklin*, Vol. 4, pp. 118–119 (Franklin to Parker, March 20, 1751).

17. Gipson, *The Great Lakes Frontier*, pp. 127–131. "For though I projected": Leonard W. Larabee, *The Papers of Benjamin Franklin*, Vol. 5, *July 1, 1753, through March 31, 1755* (New Haven, Conn.: Yale University Press, 1962), p. 454 (Franklin to Collinson, December 29, 1754).

18. Gipson, *The Great Lakes Frontier*, pp. 131–135.

19. Albert Henry Smyth, ed., *The Writings of Benjamin Franklin* (New York: Macmillan, 1907), Vol. 3, p. 242 (Franklin to Collinson, December 29, 1754).

20. Timothy J. Shannon, *Indians and Colonists at the Crossroads of Empire: The Albany Congress of 1754* (Ithaca, N.Y.: Cornell University Press, 2000), pp. 241–244.

21. Gipson, *The Great Lakes Frontier*, pp. 144–166. "I have no leaf": *Shirley Papers*, Vol. 2, pp. 95–96 (Shirley to Morris, October 21, 1754). "The assemblies did not": Walter Isaacson, *Benjamin Franklin: An American Life* (New York: Simon and Schuster, 2003), p. 161.

4 • Braddock's Roads

1. Thomas Mante, *The History of the Late War in North-America* (London: Strahan and Cadell, 1772), p. 21.

2. "Edward Braddock." *Dictionary of National Biography*, Vol. 2 (London: Oxford University Press, 1917), pp. 1061–1062. See also the standard biography of Braddock: Lee McCardell, *Ill-Starred General: Braddock of the Coldstream Guards* (Pittsburgh, Pa.: University of Pittsburgh Press, 1958).

3. T. R. Clayton, "The Duke of Newcastle, the Earl of Halifax, and the American Origins of the Seven Years' War," *Historical Journal*, Vol. 24 (1981), pp. 590–591.

4. Ibid., pp. 592–594.

5. Matthew C. Ward. *Breaking the Backcountry: The Seven Years' War in Virginia and Pennsylvania, 1754–1765* (Pittsburgh, Pa.: University of Pittsburgh Press, 2003), p. 37.

6. Lawrence Henry Gipson, *The British Empire before the American Revolution, Vol. 6, The Years of Defeat, 1754–1757* (New York: Knopf, 1946), pp. 99–103.

7. Ibid., pp. 104–116; Guy Frégault, *Canada: The War of the Conquest*, trans. Margaret M. Cameron (Toronto: Oxford University Press, 1969), pp. 90–91.

8. Ward, *Breaking the Backcountry*, p. 45.

9. Gipson, *The Years of Defeat*, p. 117.

10. Francis Jennings, *Empire of Fortune: Crowns, Colonies, and Tribes in the Seven Years War in America* (New York: Norton, 1988), p. 152.

11. "I cannot say": Ward, *Breaking the Backcountry*, p. 38. Paul E. Kopperman, *Braddock on the Monongahela* (Pittsburgh, Pa.: University of Pittsburgh Press, 1977), p. 7.

12. Walter O'Meara, *Guns at the Forks* (Englewood Cliffs, N.J.: Prentice Hall, 1965), pp. 64–72.

13. Jennings, *Empire of Fortune*, pp. 67–68. "I have my own": O'Meara. *Guns at the Forks*, pp. 126–127.

14. Kopperman, *Braddock on the Monongahela*, p. 9; for one estimate of Braddock's initial force, see Francis Parkman, *Montcalm and Wolfe: The French and Indian War (New York: Da Capo, 1995)*, p. 118.

15. Kopperman, *Braddock on the Monongahela*, pp. 10–18, 32–39.

16. Ibid., pp. 39–40, 285; for Washington's physical condition see Fred Anderson, *Crucible of War: The Seven Years' War and the Fate of Empire in British North America, 1754–1766* (New York: Knopf, 2000), p. 97.

17. Jennings, *Empire of Fortune*, pp. 153–156; Gipson, *The Years of Defeat*, pp. 70–71.

18. Jennings, *Empire of Fortune*, p. 156; Kopperman, *Braddock on the Monongahela*, p. 30.

19. "When we endeavored": W. W. Abbot, ed., *The Papers of George Washington*, Colonial Series, Vol. 1, *1748–August 1755* (Charlottesville: University Press of Virginia, 1983), pp. 339–340. Among the best analyses of the battle and contemporary accounts are Kopperman,

Braddock on the Monongahela; and Stanley Pargellis, "Braddock's Defeat," *American Historical Review*, Vol. 41 (1936), pp. 253–269.

20. Kopperman, *Braddock on the Monongahela*, pp. 91–92. For reports of French casualties see Jennings, *Empire of Fortune*, p. 158; and Frégault, *Canada: The War of the Conquest*, p. 96.

21. Pargellis, "Braddock's Defeat," pp. 264–265. Kopperman, *Braddock on the Monongahela*, "Bibliographical Essay," pp. 142–154, contains a thorough review of historical interpretations.

22. Frégault, *Canada: The War of the Conquest*, p. 97.

23. *Maryland Gazette*, August 28, 1755.

24. Parkman, *Montcalm and Wolfe*, p. 169.

25. Daniel Marston, *The French-Indian War, 1754–1760* (Oxford: Osprey, 2002), pp. 32–33, 78.

26. "Thomas Gage," *Dictionary of National Biography* (London: Macmillan, 1908), Vol. 7, pp. 795–796.

5 • "That I Can Save England"

1. Walter L. Dorn, *Competition for Empire, 1740–1763 (New York: Harper, 1940)*, p. 105.

2. Lawrence Henry Gipson, *The British Empire before the American Revolution*, Vol. 6, *The Years of Defeat, 1754–1757* (New York: Knopf, 1946), p. 402.

3. *Ibid.*

4. *Ibid.*, pp. 400–417. A. T. Mahan, *The Influence of Sea Power upon History, 1660–1783* (New York: Dover, 1987), pp. 286–287 (reprint of 1894 edition). "In this country": Voltaire, *Candide* (New York: Random House, 1930), p. 111.

5. Samuel Eliot Morison, *The Oxford History of the American People* (New York: Oxford University Press, 1965), p. 164.

6. W. H. Moreland and Atul Chandra Chatterjee, *A Short History of India*, 4th ed. (New York: McKay, 1957), pp. 265, 270–272.

7. *Pennsylvania Colonial Records*, Vol. 6, p. 513 (Morris to Shirley, July 30, 1755). "Sure I am" and "in short": *Pennsylvania Gazette*, October 30, 1755.

8. *Boston Gazette*, October 27, 1755.

9. Seymour I. Schwartz, *The French and Indian War, 1754–1763: The Imperial Struggle for North America* (New York: Castle, 1994), pp. 78–81. See also Daniel Marston, *The French-Indian War, 1754–1760*

(Oxford: Osprey, 2002), pp. 36–37; Gipson, *The Years of Defeat*, pp. 193–200.

10. Guy Frégault, *Canada: The War of the Conquest*, trans. Margaret M. Cameron (Toronto: Oxford University Press, 1969), p. 137.

11. Morison, *History of the American People*, p. 164.

12. J. C. Long, *Mr. Pitt and America's Birthright: A Biography of William Pitt, the Earl of Chatham, 1708–1778* (New York: Stokes, 1940), pp. 239–240.

13. Ibid., p. 19.

14. William B. Willcox, *A History of England*, Vol. 3, *The Age of Aristocracy, 1688–1830* (Lexington, Mass.: Heath, 1971), p. 71.

15. Long, *Mr. Pitt*, p. 242.

16. Ibid., pp. 250–253.

17. Francis Parkman, *Montcalm and Wolfe: The French and Indian War* (New York: Da Capo, 1995), p. 328.

18. Willcox, *The Age of Aristocracy*, p. 76. For more on Pitt's rise, see Fred Anderson, *Crucible of War: The Seven Years' War and the Fate of Empire in British North America, 1754–1766* (New York: Knopf, 2000), pp. 172–175; and Parkman, *Montcalm and Wolfe*, pp. 328–332.

19. Long, *Mr. Pitt*, p. 315.

6 • Massacre and Stalemate

1. Lawrence Henry Gipson, *The British Empire before the American Revolution*, Vol. 7, *The Victorious Years, 1758–1760* (New York: Knopf, 1949), p. 62.

2. William B. Willcox, *A History of England*, Vol. 3, *The Age of Aristocracy* (Lexington, Mass.: Heath, 1971), pp. 75–76.

3. A. T. Mahan, *The Influence of Sea Power upon History, 1660–1783* (New York: Dover, 1987), pp. 288–289. "Louis XV and": Francis Parkman, *Montcalm and Wolfe: The French and Indian War* (New York: Da Capo, 1995), pp. 207, 211.

4. Parkman, *Montcalm and Wolfe*, pp. 208–210.

5. Ibid., pp. 213, 216–217.

6. John A. Garraty, *The American Nation: A History of the United States* (New York: Harper and Row, 1966), pp. 98–99.

7. Guy Frégault, *Canada: The War of the Conquest*, trans. Margaret M. Cameron (Toronto: Oxford University Press, 1969), p. 137.

8. *Shirley Papers*, Vol. 2, p. 557.

9. Samuel Eliot Morison, *Oxford History of the American People* (New York: Oxford University Press, 1965), p. 164. When William Shirley left America to face his detractors in England, he took with him the best model that the English colonies had ever seen for cooperation between an English-born, crown-appointed royal governor and the sometimes recalcitrant, sometimes impetuous colonial legislatures. Shirley had been the king's man—there was no doubt about that—but he had also been keenly sensitive, too, and respectful of the differences between London and Boston. America was different, and Shirley had managed to pursue his king's imperial prerogatives without trampling on some measure of colonial self-governance in local affairs. The best evidence of this was the tribute that the Massachusetts legislature bestowed on Shirley upon his departure. "Justice as well as gratitude," the delegates avowed, "would oblige us to bear our testimony to the world that the affairs of this province have been so wisely conducted by your excellency that your name ought to be ever dear to the inhabitants." It was a far cry from what Massachusetts would be saying about its royal governor scarcely a decade hence.)

10. Gipson, *The Victorious Years*, p. 91.

11. Stanley M. Pargellis, *Lord Loudoun in North America* (New Haven, Conn.: Yale University Press, 1933), pp. 231–232; Fred Anderson, *Crucible of War: The Seven Years' War and the Fate of Empire in British North America, 1754–1766* (New York: Knopf, 2000), pp. 179–180, 183–184; Gipson, *The Victorious Years*, pp. 91–95.

12. Frégault, *Canada: The War of the Conquest*, p. 137.

13. Parkman, *Montcalm and Wolfe*, p. 264.

14. Fred Anderson, *Crucible of War*, pp. 185–186.

15. Gipson, *The Victorious Years*, p. 72.

16. Ibid., p. 81.

17. Edward P. Hamilton, ed., *Adventure in the Wilderness: The American Journals of Louis Antoine de Bougainville, 1756–1760* (Norman: University of Oklahoma Press, 1964), pp. 142–143 (July 24, 1757).

18. Anderson, *Crucible of War*, pp. 189–190.

19. Gipson, *The Victorious Years*, p. 80.

20. Ian K. Steele, *Betrayals: Fort William Henry and the "Massacre"* (New York: Oxford University Press, 1990), p. 98.

21. Wilbur R. Jacobs, *Wilderness Politics and Indian Gifts: The Northern*

Colonial Frontier, 1748–1763 (Lincoln: University of Nebraska Press, 1966), p. 177.

22. Gipson, *The Victorious Years*, p. 74.

23. "In the midst": Hamilton, *Adventure in the Wilderness*, p. 149 (July 27, 1757). "Accommodate, appease" and "thirty-three different": Anderson, *Crucible of War*, p. 189.

24. *New York Gazette*, August 8, 1757.

25. Gipson, *The Victorious Years*, pp. 80–81.

26. Jabez Fitch, Jr., *The Diary of Jabez Fitch, Jr.*, 2nd ed. (Glen Falls, N.Y.: Rogers Island Historical Association, 1968), p. 17.

27. "Does not think": Anderson, *Crucible of War*, p. 194. Hamilton, *Adventure in the Wilderness*, pp. 163, 166–167.

28. Gipson, *The Victorious Years*, p. 84.

29. Fitch, *Diary*, p. 18.

30. Parkman, *Montcalm and Wolfe*, pp. 293–294; Frye is quoted on p. 560. See also Francis Jennings, *Empire of Fortune: Crowns, Colonies, and Tribes in the Seven Years War in America* (New York: Norton, 1988), pp. 317–320; and Anderson, *Crucible of War*, pp. 195–198, for other accounts.

31. Gipson, *The Victorious Years*, p. 85.

32. Jennings, *Empire of Fortune*, pp. 317–318. Bougainville is quoted in Parkman, *Montcalm and Wolfe*, p. 251.

33. *New York Mercury*, August 22, 1757.

34. Anderson, *Crucible of War*, p. 208.

35. *Boston Evening Post*, October 24, 1757.

36. J. C. Long, *Mr. Pitt and America's Birthright: A Biography of William Pitt, the Earl of Chatham, 1708–1778* (New York: Stokes, 1940), pp. 269–270.

37. Willcox, *The Age of Aristocracy*, p. 77.

38. John Bartlett, *Bartlett's Familiar Quotations*, 13th ed. (Boston, Mass.: Little, Brown, 1955) p. 351.

7 · Fortress Atlantis

1. *New York Colonial Documents*, Vol. 10, p. 578 (Montcalm to de Maras, July 11, 1757).

2. Walter L. Dorn, *Competition for Empire, 1740–1763* (New York: Harper, 1940), pp. 357–358.

3. *New York Mercury*, August 22, 1757.

4. Dorn, *Competition for Empire*, p. 358.

5. Guy Frégault, *Canada: The War of the Conquest*, trans. Margaret M. Cameron (Toronto: Oxford University Press, 1969), p. 137.

6. "Jeffery Amherst," *Dictionary of National Biography*, Vol. 1 (New York: Macmillan, 1908), pp. 357–359; Lawrence Henry Gipson, *The British Empire before the American Revolution*, Vol. 7, *The Victorious Years* (New York: Knopf, 1949), pp. 182–183. See also J. C. Long, *Lord Jeffery Amherst: A Soldier of the King* (New York: Macmillan, 1933).

7. *Boston Evening Post*, August 4, 1755.

8. A. J. B. Johnston et al., *Louisbourg: An Eighteenth-Century Town* (Halifax: Nimbus, 1991), pp. 1–3, 6–14, 20–22, 41.

9. *Pennsylvania Gazette*, April 27, 1758.

10. Gipson, *The Victorious Years*, pp. 184–191.

11. Dorn, *Competition for Empire*, p. 358.

12. Beckles Willson, *The Life and Letters of James Wolfe* (London: William Heinemann, 1909), p. 363 (Wolfe to Sackville, May 12, 1758).

13. Ibid., p. 38 (Wolfe to his father, July 4, 1743).

14. Ibid., p. 280 (Wolfe to his mother, November 8, 1755).

15. Ibid., p. 338 (Wolfe to his father, October 24, 1757); and p. 339 (Wolfe to Rickson, November 5, 1757).

16. Ibid., p. 349 (Wolfe to Rickson, January 12, 1758).

17. J. Clarence Webster, ed., *The Journal of Jeffery Amherst: Recording the Military Career of General Amherst in America from 1758 to 1763* (Toronto: Ryerson, 1931), pp. 33–46.

18. Gipson, *The Victorious Years*, pp. 191–196.

19. Webster, *Amherst Journal*, pp. 50–51.

20. Willson, *Life and Letters of Wolfe*, pp. 384–385 (Wolfe to Walter Wolfe, July 27, 1758).

21. *Boston Gazette*, July 10, 1758 (letter dated onboard *Namur* off Louisbourg, June 9, 1758; reprinted from *Halifax Gazette*, June 24, 1758).

22. J. S. McLennan, *Louisbourg: From Its Foundation to Its Fall, 1713–1758* (Halifax: Book Room, 1979), p. 260.

23. *Boston Gazette*, July 3, 1758 (letter dated Louisbourg, June 15, 1758).

24. McLennan, *Louisbourg*, p. 233.

25. Ibid., pp. 195, 233, 234.

26. "Being doomed": Gipson, *The Victorious Years*, p. 201. See also McLennan, *Louisbourg*, pp. 242, 263, 265, 267, 301.

27. Frégault, *Canada: The War of the Conquest*, p. 219.

28. Gipson, *The Victorious Years*, pp. 197, 201–202; McLennan, *Louisbourg*, pp. 264, 266.

29. McLennan, *Louisbourg*, p. 276.

30. *Boston Gazette*, August 14, 1758 (letter dated Louisbourg, June 24, 1758).

31. Gipson, *The Victorious Years*, pp. 203–206; McLennan, *Louisbourg*, pp. 284–285.

32. *Boston Gazette*, August 21, 1758 (letter dated Gabarus Bay, July 29, 1758).

33. *Boston Gazette*, August 28, 1758 (letter dated Louisbourg, July 29, 1758).

34. *Boston Gazette*, August 14, 1758 (letter dated Louisbourg, June 24, 1758).

35. Gipson, *The Victorious Years*, p. 196.

36. Willson, *Life and Letters of Wolfe*, p. 385 (Wolfe to Walter Wolfe, July 27, 1758).

37. *Boston Evening Post*, August 28, 1758 (extract from letter of August 12, 1758).

38. McLennan, *Louisbourg*, p. 290; Frégault, *Canada: The War of the Conquest*, p. 196.

39. McLennan, *Louisbourg*, p. 311.

40. J. C. Long, *Mr. Pitt and America's Birthright: A Biography of William Pitt, the Earl of Chatham, 1708–1778* (New York: Stokes, 1940), pp. 292–295.

8 • "Till We Meet at Ticonderoga"

1. Lawrence Henry Gipson, *The British Empire before the American Revolution*, Vol. 7, *The Victorious Years* (New York: Knopf, 1949), p. 211.

2. Fred Anderson, *Crucible of War: The Seven Years' War and the Fate of Empire in British North America, 1754–1766* (New York: Knopf, 2000), pp. 225–227. Pitt's policies of giving more control over troops to colonial legislatures and promising them financial aid were a marked contrast to prior earlier directives to recruit colonials into regular regiments and essentially tax the colonial assemblies to benefit a common fund. Unwittingly, however, at the expense of winning the present war, Pitt gave them a taste of equality and fiscal reward that was to be reversed at the war's conclusion and lead to bitter dissent.

3. *Boston Gazette*, May 29, 1758.

4. *Boston Gazette*, June 19, 1758.

5. *Boston Gazette*, June 26, 1758.

6. "Makes me tremble": *New York Colonial Documents*, Vol. 10, p. 686 (Montcalm to de Moras, February 19, 1758); and p. 691 (Montcalm to de Paulmy, February 23, 1758).

7. Gipson, *The Victorious Years*, pp. 172–173.

8. *New York Colonial Documents*, Vol. 10, pp. 704–706 (Daine to de Belle Ilse, May 19, 1758).

9. Beckles Willson, *The Life and Letters of James Wolfe* (London: William Heinemann, 1909), p. 392 (Wolfe to Sackville, August 7, 1758).

10. Parkman has the most biographical information on Rogers, although it is hardly laudatory. Gipson and Anderson mention him almost in passing. Frégault mentions him not at all.

11. John R. Cuneo, *Robert Rogers of the Rangers* (New York: Oxford University Press, 1959), pp. 4, 8, 12–15, 18.

12. Robert Rogers, *Reminiscences of the French War with Robert Rogers' Journal and a Memoir of General Stark* (Freedom, N.H: Freedom Historical Society, 1988), pp. 27–31; hereafter *Rogers' Journal*. See also Cuneo, *Rogers of the Rangers*, pp. 45–49.

13. Armand Francis Lucier, *French and Indian War Notices Abstracted from Colonial Newspapers*, Vol. 2, *1756–1757* (Bowie, Md.: Heritage, 1999), pp. 74, 282.

14. Cuneo, *Rogers of the Rangers*, pp. 61–67.

15. Ibid., p. 71.

16. *Rogers' Journal*, p. 63.

17. *Boston Evening Post*, July 3, 1758 (letter dated June 12, 1758).

18. Parkman, *Montcalm and Wolfe: The French and Indian War* (New York: Da Capo, 1995), p. 357.

19. *Boston Evening Post*, July 3, 1758; Rogers, *Rogers' Journal*, pp. 65, 67.

20. *Rogers' Journal*, pp. 66–67.

21. *Boston Gazette*, July 3, 1758.

22. Francis Parkman, *Montcalm and Wolfe*, p. 358.

23. Ibid.; Cuneo, *Rogers of the Rangers*, pp. 60–61.

24. Part of the legend stems from the fact that in making his escape, Rogers threw down his coat and left it on the ground to be discovered by his pursuers. His commission papers were in a pocket, and for a time the French believed that they had killed their famous enemy. See Parkman, *Montcalm and Wolfe*, p. 312n, who is skeptical;

Cuneo, *Rogers of the Rangers*, p. 78, who pleads ignorance; and Rogers's own *Journal*, p. 52, where the story should be if it occurred.

25. Willson, *Life and Letters of Wolfe*, p. 384 (Wolfe to Walter Wolfe, July 27, 1758); *Rogers' Journal*, p. 68.

26. "A sharp fire": *Rogers' Journal*, p. 68. "So far things": Gipson, *The Victorious Years*, pp. 224–225.

27. Anderson, *Crucible of War*, p. 241.

28. Ibid., p. 243.

29. Ibid., p. 244.

30. "As hot a fire": Frederick B. Richards. *The Black Watch at Ticonderoga and Major Duncan Campbell of Inverawe* (excerpt from *Proceedings of the New York State Historical Association*, Vol. 10, printed for the Ticonderoga Museum Library, n. d.), p. 24, quoting Captain James Murray to his brother, July 19, 1758; see also pp. 27 and 52 for casualty figures. For other accounts of the battle, see Parkman, *Montcalm and Wolfe*, pp. 361–368; and Gipson, *The Victorious Years*, pp. 225–231.

31. This version is from Parkman, *Montcalm and Wolfe*, pp. 561–563; but the most famous telling may be Robert Louis Stevenson's poem, "Ticonderoga: A Legend of the West Highlands."

32. Gertrude Selwyn Kimball, *Correspondence of William Pitt When Secretary of State* (New York: Macmillan, 1906), Vol. 1, p. 300; hereafter *Pitt Correspondence*.

33. Parkman, *Montcalm and Wolfe*, p. 561; Guy Frégault, *Canada: The War of the Conquest*, trans. Margaret M. Cameron (Toronto: Oxford University Press, 1969), p. 221. According to one source, this was the most deadly battle fought on American soil until the Civil War, with the exception of the battle of Long Island during the American Revolution; see Edward P. Hamilton, *Fort Ticonderoga: Key to a Continent* (Boston, Mass.: Little, Brown, 1964), p. 85.

34. Parkman, *Montcalm and Wolfe*, p. 368.

35. Armand Francis Lucier, *French and Indian War Notices Abstracted from Colonial Newspapers*, Vol. 3, *January 1, 1758–September 17, 1759* (Bowie, Md.: Heritage, 1999), p. 162.

36. Anderson, *Crucible of War*, p. 248.

9 • The Bateau Man

1. J. C. Long, *Lord Jeffery Amherst: A Soldier of the King* (New York: Macmillan, 1933), p. 76.

2. "The unlucky accident": Beckles Willson, *The Life and Letters of Wolfe* (London: William Heinemann, 1909), p. 391 (Wolfe to Sackville, August 7, 1758). *Rogers' Journal*, pp. 71–72. Francis Parkman, *Montcalm and Wolfe: The French and Indian War* (New York: Da Capo, 1995), pp. 375–377.

3. Harrison Bird, *Battle for a Continent: The French and Indian War, 1754–1763* (New York: Oxford University Press, 1965), p. 190. Contemporary accounts spell bateau, singular, and bateaux, plural, in a variety of ways—some quite comical—but all spellings have been corrected in these quotations.

4. Fred Anderson, *Crucible of War: The Seven Years' War and the Fate of Empire in British North America, 1754–1766* (New York: Knopf, 2000), pp. 259, 778n; Lawrence Henry Gipson, *The British Empire before the American Revolution*, Vol. 7, *The Victorious Years* (New York: Knopf, 1949), p. 236.

5. *Shirley Papers*, Vol. 2, pp. 240–241 (Bradstreet to Shirley, August 17, 1755).

6. Ibid., pp. 442–443 (Shirley to Fox, May 7, 1756).

7. Ibid., p. 580 (Morris to Sharpe, October 8, 1756).

8. Gipson, *The Victorious Years*, p. 237.

9. Ibid., pp. 238–239.

10. Douglas Edward Leach, *The Northern Colonial Frontier, 1607–1763* (New York: Holt, Rinehart, and Winston, 1966), p. 105.

11. Gipson, *The Victorious Years*, pp. 242–243. Fifty-five years later, after a great debate about the strategic importance and defensibility of the same location—then Kingston, Upper Canada—American troops during the war of 1812 sailed across Lake Ontario and attacked York (now Toronto) in a similar raid. It resulted in little more than the burning of the government buildings there, but this led to a thirst for retaliation that was expressed in the burning of buildings in Washington, D. C., two years later. Kingston then, like Fort Frontenac fifty-five years earlier, was the key to severing Canada. Bradstreet had recognized it.

12. *Boston Gazette*, August 28, 1758.

13. *Boston Gazette*, September 11, 1758.

14. William G. Godfrey, *Pursuit of Profit and Preferment in Colonial North America: John Bradstreet's Quest* (Waterloo, Ont.: Wilfrid/Laurier University Press, 1982), pp. 128–131; Gipson, *The Victorious Years*, pp. 244–246; Anderson, *Crucible of War*, pp. 261–264. "Being so

near": *New York Mercury*, September 18, 1758 (letter dated Oswego, August 30, 1758).

15. "It had in it": *New York Mercury*, September 18, 1758 (letter dated Oswego, August 30, 1758. "2000 barrels": *Boston Gazette*, September 18, 1758 (letter dated Oswego, August 30, 1758).

16. *Boston Gazette*, September 18, 1758 (letter dated Oswego, August 30, 1758).

17. Gipson, *The Victorious Years*, p. 246.

18. Willson, *Life and Letters of Wolfe*, pp. 369, 403 (Wolfe to Sackville, May 24, 1758; Wolfe to Rickson, December 1, 1758).

19. Edward P. Hamilton, ed., *Adventure in the Wilderness: The American Journals of Louis Antoine de Bougainville, 1756–1760* (Norman: University of Oklahoma Press, 1964), p. 273.

20. Gipson, *The Victorious Years*, pp. 218, 220.

10 • Braddock's Roads Again

1. Walter O'Meara, *Guns at the Forks* (Englewood Cliffs, N.J.: Prentice Hall, 1965) p. 158.

2. Lawrence Henry Gipson, *The British Empire before the American Revolution*, Vol 7, *The Victorious Years* (New York: Knopf, 1949), pp. 247–250. "Utter stranger": *Pitt Correspondence*, Vol. 1, p. 252 (Abercromby to Pitt, May 22, 1758).

3. Gipson, *The Victorious Years*, pp. 258–259. "All suspected places": S. K. Stevens, Donald H. Kent, and Autumn L. Leonard, eds., *The Papers of Henry Bouquet*, Vol. 1, *December 11, 1755–May 31, 1758* (Harrisburg: Pennsylvania Historical and Museum Commission, 1972), pp. 52–53.

4. "The villainous behavior": W. W. Abbot, ed., *The Papers of George Washington*, Colonial Series, Vol. 5, *October 1757–September 1758* (Charlottesville: University Press of Virginia, 1988), p. 12 (Washington to Dinwiddie, October 9, 1757). "A few of their": Alfred Proctor James, ed., *The Writings of General John Forbes Relating to His Service in North America* (Menasha, Wis.: The Collegiate, 1938), p. 205 (Forbes to Pitt, September 6, 1758).

5. James, *Writings of Forbes*, p. 129 (Forbes to Bouquet, July 6, 1758).

6. Gipson, *The Victorious Years*, pp. 261–262.

7. "A new way": Abbot, *Papers of George Washington*, Colonial Series, Vol. 5, p. 360 (Washington to Halkett, August 2, 1758). "The Vir-

ginians are": Gipson, *The Victorious Years*, p. 264n. The determination of the young Washington to press for the Virginia route and the exasperation of the proper Bouquet make for fascinating reading in Washington's papers. See in particular Washington to Bouquet, August 2, 1758.

8. "Going into Braddock's": James, *Writings of Forbes*, p. 129 (Forbes to Bouquet, July 6, 1758). "Send me": O'Meara, *Guns at the Forks*, p. 195.

9. Abbot, *Papers of George Washington*, Colonial Series, Vol. 5, p. 365 (Bouquet to Washington, August 3, 1758).

10. Gipson, *The Victorious Years*, p. 267.

11. Guy Frégault, *Canada: The War of the Conquest*, trans. Margaret M. Cameron (Toronto: Oxford University Press, 1969), p. 99.

12. Gipson, *The Victorious Years*, pp. 271–273.

13. S. K. Stevens, Donald H. Kent, and Autumn L. Leonard, eds., *The Papers of Henry Bouquet*, Vol. 2, *The Forbes Expedition* (Harrisburg: Pennsylvania Historical and Museum Commission, 1951), pp. 502–504, 518–521.

14. Gipson, *The Victorious Years*, pp. 273–274.

15. Fred Anderson, *Crucible of War: The Seven Years' War and the Fate of Empire in British North America, 1754–1766* (New York: Knopf, 2000), pp. 267–271, 779n.

16. Matthew C. Ward, *Breaking the Backcountry: The Seven Years' War in Virginia and Pennsylvania, 1754–1765* (Pittsburgh, Pa.: University of Pittsburgh Press, 2003), pp. 178–182; Anderson, *Crucible of War*, pp. 275–278.

17. W. W. Abbot, ed., *The Papers of George Washington*, Colonial Series, Vol. 6, *September 1758–December 1760* (Charlottesville: University Press of Virginia, 1988), p. 99 (Washington to Fauquier, October 30, 1758).

18. Gipson, *The Victorious Years*, p. 282; Anderson, *Crucible of War*, pp. 282, 781n.

19. Frégault, *Canada: The War of the Conquest*, p. 224.

20. "A heavy firing": *Pennsylvania Gazette*, December 7, 1758. "Monsieurs did not stay": *New York Gazette*, December 18, 1758.

21. "After much fatigue": *Pennsylvania Gazette*, December 28, 1758. "A long row": Gipson, *The Victorious Years*, p. 283.

22. *Pennsylvania Gazette*, December 14, 1758.

23. "A vast country": *Pennsylvania Gazette*, December 28, 1758. "Blessed be God": *Pennsylvania Gazette*, December 14, 1758.

24. James, *Writings of Forbes*, p. 283 (Forbes to Amherst, January 26, 1759; see also postscript in Forbes to Amherst, January 18, 1759).

11 • Caribbean Gambit

1. Walter L. Dorn, *Competition for Empire, 1740–1763* (New York: Harper, 1940), map following p. 364; Lawrence Henry Gipson, *The British Empire before the American Revolution*, Vol. 5, *The Great Lakes Frontier, Canada, the West Indies, India, 1748–1754* (New York: Knopf, 1942), pp. 211, 215–216.

2. Lawrence Henry Gipson, *The British Empire before the American Revolution*, Vol. 8, *The Culmination, 1760–1763.* (New York: Knopf, 1953), pp. 65–67.

3. Ibid., pp. 67–71.

4. Ibid., pp. 72–74.

5. Ibid., p. 76.

6. A. T. Mahan, *The Influence of Sea Power upon History, 1660–1783* (New York: Dover, 1987), p. 291.

7. *Boston Evening Post*, August 28, 1758; reprinted from *South Carolina Gazette*, July 21, 1758.

8. *New York Mercury*, September 11, 1758 (dispatches dated Saint Augustine, August 25, 1758).

9. Given overall British losses, this number seems inflated, but it is recounted in Gipson, *The Culmination*, p. 83; Henry C. Wilkinson, *Bermuda in the Old Empire* (Oxford: Oxford University Press, 1950), p. 232; Mahan, *The Influence of Sea Power*, p. 314; and Dorn, *Competition for Empire*, p. 362, among others.

10. Gipson, *The Culmination*, pp. 83–84.

11. *Pennsylvania Gazette*, November 9, 1758 (dispatches dated Saint Pierre, June 15, 1758).

12. Gipson, *The Culmination*, p. 86.

13. J. C. Long, *Mr. Pitt and America's Birthright: A Biography of William Pitt, the Earl of Chatham, 1708–1778* (New York: Stokes, 1940), p. 299.

14. Fred Anderson, *Crucible of War: The Seven Years' War and the Fate of Empire in British North America, 1754–1766* (New York: Knopf, 2000), p. 306; Long, *Pitt*, p. 299.

15. Gipson, *The Culmination*, p. 85n.

16. Marshall Smelser, "The Insular Campaign of 1759: Martinique," *American Neptune*, Vol. 6, No. 4 (October 1946), pp. 291–293.

17. Ibid. pp. 294–297. "The Highlands of Scotland": Gipson, *The Culmination*, p. 91.

18. Smelser, "Martinique," pp. 298–300.

19. Marshall Smelser, "The Insular Campaign of 1759: Guadeloupe," *American Neptune*, Vol. 7, No. 1 (January 1947), pp. 21–23. "Contrary to my wishes": *Pitt Correspondence*, Vol. 2, p. 30 (Moore to Pitt, January 30, 1759).

20. *Pennsylvania Gazette*, March 8, 1759.

21. Smelser, "Guadeloupe," pp. 24–26; Gipson, *The Culmination*, pp. 100–101.

22. Smelser, "Guadeloupe," pp. 27–30; Gipson, *The Culmination*, pp. 102–105.

23. Anderson, *Crucible of War*, p. 315.

12 · Falling Dominoes

1. J. C. Long, *Lord Jeffery Amherst: A Soldier of the King* (New York: Macmillan, 1933), pp. 76–77.

2. *Boston Gazette*, October 16, 1758.

3. Long, *Amherst*, pp. 78– 80, 82.

4. Guy Frégault, *Canada: The War of the Conquest*, trans. Margaret M. Cameron (Toronto: Oxford University Press, 1969), p. 161.

5. Ibid., pp. 227–229.

6. *Pitt Correspondence*, Vol. 2, p. 9 (Amherst to Pitt, January 18, 1759).

7. *Pitt Correspondence*, Vol. 1, p. 438 (Pitt to Amherst, December 29, 1758).

8. Long, *Amherst*, p. 84.

9. Lawrence Henry Gipson, *The British Empire before the American Revolution* Vol. 7, *The Victorious Years* (New York: Knopf, 1949), p. 334. "To enable him": *Boston Gazette*, July 23, 1759.

10. *Pennsylvania Gazette*, May 3, 1759 (letter dated Fort Ligonier, April 17, 1759).

11. "This morning": *New York Mercury*, June 11, 1759 (letter dated Winchester, May 28, 1759). "We cannot find": *Maryland Gazette*, June 14, 1759.

12. Gipson, *The Victorious Years*, pp. 344–345.

13. James Sullivan, *The Papers of Sir William Johnson* (Albany: University of the State of New York, 1921), Vol. 3, pp. 19–20 (Johnson to Amherst, February 16, 1759).

14. Francis Jennings, *Empire of Fortune: Crowns, Colonies, and Tribes in the Seven Years War in America* (New York: Norton, 1988), pp. 415–416.

15. Gipson, *The Victorious Years*, pp. 345, 347–348.

16. Robert West Howard, *Thundergate: The Forts of Niagara* (Englewood, N.J: Prentice Hall, 1968), pp. 90–91.

17. Fred Anderson, *Crucible of War: The Seven Years' War and the Fate of Empire in British North America, 1754–1766* (New York: Knopf, 2000), pp. 335–336.

18. Gipson, *The Victorious Years*, p. 351.

19. Howard, *Thundergate*, pp. 94–95; the number of cannonballs fired is in the *Pennsylvania Gazette*, August 2, 1759, from a letter dated Niagara, July 16, 1759.

20. "Floating island": *New York Colonial Documents*, Vol. 10, p. 986. Jennings, *Empire of Fortune*, pp. 417–418.

21. "The men received": Stephen Brumwell, *Redcoats: The British Soldier and War in the Americas, 1755–1763* (Cambridge: Cambridge University Press, 2002), p. 253. "As I hear": Gipson, *The Victorious Years*, pp. 352–355. Anderson, *Crucible of War*, pp. 337–338.

22. Armand Francis Lucier, *French and Indian War Notices Abstracted from Colonial Newspapers*, Vol. 3, *January 1, 1758–September 17, 1759* (Bowie, Md.: Heritage, 1999), p. 238.

23. Francis Parkman, *Montcalm and Wolfe: The French and Indian War* (New York: Da Capo, 1995), p. 443; Anderson, *Crucible of War*, pp. 340–341.

24. Long, *Amherst*, pp. 100–101. (The biblical reference is Judges 7:25 and 8:21.)

25. Long, *Amherst*, pp. 105, 109; Anderson, *Crucible of War*, pp. 340, 342–343.

13 • Battle for a Continent—Or Is It?

1. Beckles Willson, *The Life and Letters of James Wolfe* (London: William Heinemann, 1909), p. 396 (Wolfe to his father, August 21, 1758); p. 397 (Wolfe to Amherst, September 30, 1758).

2. Ibid., p. 400.

3. Ibid., pp. 400–401 (Wolfe to Pitt, November 22, 1758).

4. Ibid., p. 405 (Wolfe to Parr, December 6, 1758).

5. Ibid., pp. 93, 406 (Wolfe to Rickson, April 2, 1749).

6. Ibid., pp. 415–417.

7. Ibid., p. 427 (Wolfe to his uncle Walter Wolfe, May 19, 1759).

8. Ibid., p. 388 (Wolfe to Sackville, July 30, 1758).

9. Ibid., pp. 413–414. See also Christopher Hibbert, *Wolfe at Quebec: The Man Who Won the French and Indian War* (New York: Cooper Square, 1999), pp. 32–34 (originally published 1959); Stuart Reid, *Wolfe: The Career of General James Wolfe from Culloden to Quebec* (Rockville Center, N.Y.: Sarpedon, 2000), pp. 164–165.

10. Robert Leckie, *A Few Acres of Snow: The Saga of the French and Indian Wars* (New York: Wiley, 1999), p. 119.

11. Fred Anderson, *Crucible of War: The Seven Years' War and the Fate of Empire in British North America, 1754–1766* (New York: Knopf, 2000), p. 345.

12. Armand Francis Lucier, *French and Indian War Notices Abstracted from Colonial Newspapers*, Vol. 3, *January 1, 1758–September 17, 1759* (Bowie, Md.: Heritage, 1999), pp. 262–263 (letter dated Quebec, April 30, 1759, printed in the *Halifax Gazette* of June 30, 1759).

13. *Pitt Correspondence*, Vol. 2, p. 8, (Amherst to Pitt, January 18, 1759).

14. Reid, *Wolfe*, p. 168.

15. Lawrence Henry Gipson, *The British Empire before the American Revolution*, Vol. 7, *The Victorious Years* (New York: Knopf, 1949), pp. 376–377.

16. Ibid., pp. 379–380. "The enemy have": Willson, *Life and Letters of Wolfe*, p. 436.

17. *Boston Gazette*, September 10, 1759 (letter dated Point Lévis, July 29, 1759).

18. Willson, *Life and Letters of Wolfe*, pp. 457–459.

19. Gipson, *The Victorious Years*, p. 396.

20. Ibid., pp. 401–404; Anderson, *Crucible of War*, pp. 342, 383.

21. "The public service": Willson, *Life and Letters of Wolfe*, p. 466. "General Wolfe's health": Gipson, *The Victorious Years*, p. 405. Anderson, *Crucible of War*, p. 351.

22. Gipson, *The Victorious Years*, pp. 406–407. "My ill-state": Willson, *Life and Letters of Wolfe*, p. 475.

23. Gipson, *The Victorious Years*, pp. 407–409.

24. Ibid., pp. 409–410, 412; Anderson, *Crucible of War*, pp. 351–352.

25. Willson, *Life and Letters of Wolfe*, pp. 482–486.

26. Gipson, *The Victorious Years*, pp. 412–414.

27. Ibid., pp. 415–417; Willson, *Life and Letters of Wolfe*, pp. 487–488; Anderson, *Crucible of War*, pp. 354–355.

28. Anderson, *Crucible of War*, pp. 359–363.

29. Gipson, *The Victorious Years*, pp. 418–419.

30. Ibid., pp. 421, 423.

31. Anderson, *Crucible of War*, pp. 365, 368. "The English hold" Guy Frégault, *Canada: The War of the Conquest*, trans. Margaret M. Cameron (Toronto: Oxford University Press, 1969), p. 257.

32. "Carthage may boast": Frégault, *Canada: The War of the Conquest*, pp. 261–262. "The Maple Leaf Forever," in *Children's Very First Piano Pieces* (New York: Edward Schuberth, 1947), p. 60.

14 • The Making of a Legend

1. Lawrence Henry Gipson, *The British Empire before the American Revolution*, Vol. 7, *The Victorious Years* (New York: Knopf, 1949), pp. 364–365; Burt Garfield Loescher, *The History of Rogers' Rangers: The St. Francis Raid* (Bowie, Md.: Heritage, 2002), pp. xix, 155–156.

2. *Rogers' Journal*, p. 85.

3. "He could not have": Gipson, *The Victorious Years*, p. 365. "Some 600 scalps": *Rogers' Journal*, p. 90. Francis Jennings, *Empire of Fortune: Crowns, Colonies, and Tribes in the Seven Years War in America* (New York: Norton, 1988), p. 189.

4. Burt Garfield Loescher, *Genesis: Rogers' Rangers, The First Green Berets* (Bowie, Md.: Heritage, 2000), p. 57; Loescher, *St. Francis*, pp. 4–7.

5. Loescher, *Genesis*, pp. 57–58. "We now determined": *Rogers' Journal*, p. 89.

6. Loescher, *St. Francis*, pp. 29, 35, 37, 39, 42; *Rogers' Journal* pp. 85–86.

7. Loescher, *St. Francis*, pp. 40, 46, 52–53.

8. "As long as you": John R. Cuneo, *Robert Rogers of the Rangers* (New York: Oxford University Press, 1959), p. 105.

9. Loescher, *St. Francis*, pp. 83, 87, 88–90.

10. Ibid., pp. 94, 96, 197, 199–200.

11. Ibid., p. 180.

15 • Deciding the Fate

1. "We are masters": Daniel Marston, *The French-Indian War, 1754–1760* (Oxford: Osprey, 2002), p. 64. "A severe winter": Lawrence Henry Gipson, *The British Empire before the American Revolution*, Vol. 7, *The Victorious Years* (New York: Knopf, 1949), p. 431.

2. "Totally unfit": *Pitt Correspondence*, Vol. 2, pp. 292 (Murray to Pitt, May 25, 1760). "The English hold": Edward P. Hamilton, ed., *Adventure in the Wilderness: The American Journal of Louis Antoine de Bougainville, 1756–1760* (Norman: University of Oklahoma Press, 1964), p. 321.

3. Gipson, *The Victorious Years*, pp. 429, 435–436.

4. Ibid., pp. 437–438.

5. Pitt, *Correspondence*, Vol. 2, p. 293, (Murray to Pitt, May 25, 1760).

6. "Our cannon were": Marston, *The French-Indian War*, p. 67; Gipson, *The Victorious Years*, pp. 339–340.

7. Fred Anderson, *Crucible of War: The Seven Years' War and the Fate of Empire in British North America, 1754–1766* (New York: Knopf, 2000), p. 394.

8. Walter L. Dorn, *Competition for Empire, 1740–1763* (New York: Harper, 1940), p. 355; Lawrence Henry Gipson, *The British Empire before the American Revolution*, Vol. 8, *The Culmination, 1760–1763* (New York: Knopf, 1953), pp. 4–5.

9. A. T. Mahan, *The Influence of Sea Power upon History, 1660–1783* (New York: Dover, 1987), pp. 298–299; Gipson, *The Culmination*, pp. 12–15.

10. Gipson, *The Culmination*, pp. 18–21.

11. "I had no ground": J. C. Long, *Mr. Pitt and America's Birthright: A Biography of William Pitt, the Earl of Chatham, 1708–1778* (New York: Stokes, 1940), p. 320. "General chase": Anderson, *Crucible of War*, p. 382. "Had we but two": Julian S. Corbett, *England in the Seven Years War: A Study in Combined Strategy* (London: Longmans Green, 1918), Vol. 2, p. 69.

12. Gipson, *The Culmination*, p. 24.

13. Mahan, *The Influence of Sea Power*, p. 304.

14. Guy Frégault, *Canada: The War of the Conquest*, trans. Margaret M. Cameron (Toronto: Oxford University Press, 1969), p. 273.

16 • Montreal to Michilimackinac

1. Guy Frégault, *Canada: The War of the Conquest*, trans. Margaret M. Cameron (Toronto: Oxford University Press, 1969), p. 279.

2. *Pitt Correspondence*, Vol. 2, pp. 288–289 (Amherst to Pitt, May 19, 1760). Troop numbers and "made it impossible": Frégault, *Canada: The War of the Conquest*, p. 282.

3. "Easiest": Frégault, *Canada: The War of the Conquest*, p. 283. "The St.

Lawrence": Lawrence Henry Gipson, *The British Empire before the American Revolution*, Vol. 7, *The Victorious Years* (New York: Knopf, 1949), p. 458.

4. Frégault: *Canada: The War of the Conquest*, pp. 282, 285; Fred Anderson, *Crucible of War: The Seven Years' War and the Fate of Empire in British North America, 1754–1766* (New York: Knopf, 2000), pp. 398, 401–402.

5. Gipson, *The Victorious Years*, p. 463.

6. J. C. Long, *Lord Jeffery Amherst: A Soldier of the King* (New York: Macmillan, 1933), p. 133.

7. Gipson, *The Victorious Years*, pp. 463–466. "The infamous part": Anderson, *Crucible of War*, p. 408.

8. "Without having fired": Frégault, *Canada: The War of the Conquest*, p. 288. "After all, sire": This story and the phrase "a few acres of snow" are the basis for the title of Robert Leckie's narrative of the colonial wars in North America; Voltaire used the phrase in *Candide* (New York: Random House, 1930), which in the edition used here is in Chapter 23, p. 110 .

9. *Pitt Correspondence*, Vol. 1, pp. 387–392 (Lyttelton to Pitt, November 4, 1758).

10. Gipson, *The Victorious Years*, pp. 290–292, 292n.

11. Howard H. Peckham, *The Colonial Wars, 1689–1762* (Chicago, Ill.: University of Chicago Press, 1964), p. 202.

12. Tom Hatley, *The Dividing Paths: Cherokees and South Carolinians through the Revolutionary Era* (New York: Oxford University Press, 1995), pp. 122–130.

13. Ibid., p. 132.

14. Ibid., pp. 133, 138–139; Hugh F. Rankin, *Francis Marion: The Swamp Fox* (New York: Crowell, 1973), pp. 5–6; Anderson, *Crucible of War*, pp. 466–467.

15. James Sullivan, *The Papers of Sir William Johnson*, (Albany: University of the State of New York, 1921), Vol. 3, pp. 514–516 (Amherst to Johnson, August 9, 1761).

16. *Boston Evening Post*, October 13, 1760.

17. *Rogers' Journal*, pp. 111–113, 121–122; John R. Cuneo, *Robert Rogers of the Rangers* (New York: Oxford University Press, 1959), pp. 130–134, 292n. See also Francis Parkman, *The Conspiracy of Pontiac*. Parkman describes the Ottawa chief as meeting with Rogers at the mouth of the Cuyahoga, but Cuneo is skeptical of this.

18. "No French officer": *Rogers' Journal*, p. 125; Cuneo, *Rogers of the Rangers*, pp. 135–137.

19. "Everybody here": Cuneo, *Rogers of the Rangers*, p. 138. "To our great": *Rogers' Journal*, p. 129.

20. Cuneo, *Rogers of the Rangers*, pp. 132, 139–141.

17 • Martinique to Manila

1. Lawrence Henry Gipson, *The British Empire before the American Revolution*, Vol. 8, *The Culmination, 1760–1763* (New York: Knopf, 1953), pp. 245–250; Charles Petrie, *King Charles III of Spain: An Enlightened Despot* (London: Constable, 1971), pp. 49–51.

2. J. C. Long, *Mr. Pitt and America's Birthright: A Biography of William Pitt, the Earl of Chatham, 1708–1778* (New York: Stokes, 1940), pp. 332, 337, 342–349, 370–372; Fred Anderson, *Crucible of War: The Seven Years' War and the Fate of Empire in British North America, 1754–1766* (New York: Knopf, 2000), pp. 478, 801n.

3. Gipson, *The Culmination*, pp. 222–224.

4. Howard H. Peckham, *The Colonial Wars, 1689–1762* (Chicago, Ill.: University of Chicago Press, 1946), pp. 206–207.

5. Walter L. Dorn, *Competition for Empire, 1740–1763* (New York: Harper, 1940), pp. 367–369; A. T. Mahan, *The Influence of Sea Power upon History, 1660–1783* (New York: Dover, 1987), pp. 307–310. For a complete discussion of the Seven Years' War in India, see Gipson, *The Culmination*, pp. 108–171.

6. Gipson, *The Culmination*, pp. 185–196; Anderson, *Crucible of War*, p. 490.

7. Richard Pares, *War and Trade in the West Indies, 1739–1763* (Oxford: Clarendon, 1936), pp. 590–593; Mahan, *The Influence of Sea Power*, p. 315; Anderson, *Crucible of War*, pp. 498–501; Gipson, *The Culmination*, pp. 264–286.

8. Anderson, *Crucible of War*, pp. 515–517. "Such a banditti": Gipson, *The Culmination*, p. 277.

18 • Scratch of a Pen

1. "Half a continent": Francis Parkman, *Montcalm and Wolfe: The French and Indian War* (New York: Da Capo, 1995), p. 526. "At the end": A. T. Mahan, *The Influence of Sea Power upon History, 1660–1783* (New York: Dover, 1987), p. 291.

2. Ronald W. Clark, *Benjamin Franklin: A Biography* (New York: Random House, 1983), p. 157.

3. Richard Pares, *War and Trade in the West Indies, 1739–1763* (Oxford: Clarendon, 1936), pp. 219, 224.

4. Walter L. Dorn, *Competition for Empire, 1740–1763* (New York: Harper, 1940), pp. 375–376.

5. Walter L. Dorn, "Frederic the Great and Lord Bute," *Journal of Modern History*, Vol. 1, No. 4 (December 1929), pp. 534, 557–558.

6. Dorn, *Competition for Empire*, pp. 376–378.

7. Lawrence Henry Gipson, *The British Empire before the American Revolution*, Vol. 8, *The Culmination*, pp. 309–311. "There appears": Leonard W. Larabee, *The Papers of Benjamin Franklin*, Vol. 10, *January 1, 1762, through December 31, 1763* (New Haven, Conn.: Yale University Press, 1966), p. 215 (Franklin to Jackson, March 8, 1763).

8. Dorn, *Competition for Empire*, p. 378.

9. Gipson, *The Culmination*, p. 308.

10. Ibid., pp. 309, 311–312.

11. John A. Garraty, *The American Nation: A History of the United States* (New York: Harper and Row, 1966), p. 76.

19 • A Matter Unresolved

1. Lawrence Henry Gipson, *The British Empire before the American Revolution*, Vol. 9, *The Triumphant Empire: New Responsibilities within the Enlarged Empire, 1763–1766* (New York: Knopf, 1956), pp. 51–52; Fred Anderson, *Crucible of War: The Seven Years' War and the Fate of Empire in British North America, 1754–1766* (New York: Knopf, 2000), pp. 565–568.

2. Francis Parkman, *The Conspiracy of Pontiac and the Indian War after the Conquest of Canada* (New York: Dutton, 1908, reprint); Wilbur R. Jacobs, "Was the Pontiac Uprising a Conspiracy? *Ohio State Archaeological and Historical Quarterly*, Vol. 59, No. 1 (January 1959), pp. 30–34. For Neolin's vision, see Gregory Evans Dowd, *A Spirited Resistance: The North American Indian Struggle for Unity, 1745–1815* (Baltimore, Md.: Johns Hopkins University Press, 1992), p. 33. "Long accustomed": Wilbur R. Jacobs, *Wilderness Politics and Indian Gifts: The Northern Colonial Frontier, 1748–1763* (Lincoln: University of Nebraska Press, 1966), pp. 161, 185.

3. Howard H. Peckham, *Pontiac and the Indian Uprising* (Chicago, Ill.:

University of Chicago Press, 1961), pp. 15–18, 43–44, 47–48, 59–62.

4. Ibid., pp. 108–109, 154, 159–170.

5. Ibid., pp. 76–78, 117, 121–127, 130–135. See also Helen Humphrey, "The Identity of Major Gladwin's Informant," *Mississippi Valley Historical Review*, Vol. 21, No. 2 (September 1934), pp. 147–162.

6. Peckham, *Pontiac and the Indian Uprising*, pp. 200–210. See also John R. Cuneo, *Robert Rogers of the Rangers* (New York: Oxford University Press, 1959), pp. 164–167.

7. Gipson, *New Responsibilities*, pp. 109–110.

8. "Could it not be": Peckham, *Pontiac and the Indian Uprising*, pp. 226–227. See also Matthew C. Ward, *Breaking the Backcountry: The Seven Years' War in Virginia and Pennsylvania, 1754–1765* (Pittsburgh, Pa.: University of Pittsburgh Press, 2003), pp. 228–229; Gipson, *New Responsibilities*, p. 108. For distribution of blankets at Fort Pitt, see Francis Jennings, *Empire of Fortune: Crowns, Colonies, and Tribes in the Seven Years' War in America* (New York: Norton, 1988), pp. 447–448.

9. Gipson, *New Responsibilities*, p. 90.

10. Ibid., pp. 111–112; Ward, *Breaking the Backcountry*, p. 229.

11. Ward, *Breaking the Backcountry*, p. 233.

12. "We have visibly": Peckham, *Pontiac and the Indian Uprising*, p. 101; Gipson, *New Responsibilities*, pp. 117–118, 124–126. See also William G. Godfrey, *Pursuit of Profit and Preferment in Colonial North America: John Bradstreet's Quest* (Waterloo, Ont.: Wilfrid Laurier University Press, 1982), pp. 196–232.

13. J. C. Long, *Lord Jeffery Amherst: A Soldier of the King* (New York: Macmillan, 1933), pp. 188–189, 193. "A new king": Peckham, *Pontiac and the Indian Uprising*, p. 242.

14. Long, *Amherst*, pp. 189, 237–238.

20 · Prelude to Revolution

1. Fred Anderson, *Crucible of War: The Seven Years' War and the Fate of Empire in British North America, 1754–1766* (New York: Knopf, 2000), p. 604.

2. John A. Garraty, *The American Nation: A History of the United States* (New York: Harper and Row, 1966), pp. 80–82, 102. Paying certain taxes and other levies to the crown was nothing new for the colonies,

but heretofore this had generally been done through the colonial legislatures, as was the case with many of the appropriations during the recent war. The crown asked and the legislatures usually obliged. Popular voting for representatives to these assemblies was generally limited to white males who owned property. But the assemblies were definitely viewed as representative bodies, and their expenditures to support the crown's mission were seen very differently from direct taxes levied by Parliament, where no colonial was present to speak for or against them.

3. L. H. Butterfield, ed., *Diary and Autobiography of John Adams*, Vol. 1, *Diary 1755–1770* (Cambridge, Mass.: Belknap Press of Harvard University Press, 1961), p. 284.

4. Anderson, *Crucible of War*, pp. 700–701.

5. *Ibid*, pp. 648–650; John C. Miller, *Origins of the American Revolution* (Boston: Little, Brown, 1943), pp. 237–240.

6. "Americans are the sons": J. C. Long, *Mr. Pitt and America's Birthright: A Biography of William Pitt, the Earl of Chatham, 1708–1778* (New York: Stokes, 1940), p. 439; see also pp. 461–463. "Ungrateful brats": Garraty, *The American Nation*, p. 83.

7. Francis Jennings, *Empire of Fortune: Crowns, Colonies, and Tribes in the Seven Years War in America* (New York: Norton, 1988), pp. 463, 466–467. The thirteen colonies would finally give up their western land claims as part of the debate over the ratification of the Articles of Confederation.

8. John A. Schutz, *William Shirley: King's Governor of Massachusetts* (Chapel Hill: University of North Carolina Press, 1961), p. 265.

9. David Lavender, *The Fist in the Wilderness* (Garden City, N.Y.: Doubleday, 1964), pp. 14–15; John R. Cuneo, *Robert Rogers of the Rangers* (New York: Oxford University Press, 1959), pp. 266–267, 275–278.

10. Howard H. Peckham, *Pontiac and the Indian Uprising* (Chicago, Ill.: University of Chicago Press, 1961), pp. 261–262, 265–266, 290, 297, 311.

11. Burt Garfield Loescher, *The History of Rogers' Rangers: The St. Francis Raid* (Bowie, Md.: Heritage, 2002), pp. 215–216.

12. Long, *Mr. Pitt*, pp. 494, 507, 533.

BIBLIOGRAPHY

Books

Anderson, Fred. *Crucible of War: The Seven Years' War and the Fate of Empire in British North America, 1754–1766*. New York: Alfred A. Knopf, 2000.

Asprey, Robert B. *War in the Shadows: The Guerrilla in History*. New York: Doubleday, 1975.

Bird, Harrison. *Battle for a Continent: The French and Indian War, 1754–1763*. New York: Oxford University Press, 1965.

———. *Navies in the Mountains: The Battles on the Waters of Lake Champlain and Lake George, 1609–1814*. New York: Oxford University Press, 1962.

Bowen, Catherine Drinker. *The Most Dangerous Man in America*. Boston: Little, Brown, 1974.

Brumwell, Stephen. *Redcoats: The British Soldier and War in the Americas, 1755–1763*. Cambridge: Cambridge University Press, 2002.

Clark, Ronald W. *Benjamin Franklin: A Biography*. New York: Random House, 1983.

Clark, Thomas D. *Frontier America: The Story of the Westward Movement*. Second edition. New York: Charles Scribner's Sons, 1969.

Connell, Brian. *The Savage Years*. New York: Harper and Brothers, 1959.

Cooper, James Fenimore. *The Last of the Mohicans*. New York: New American Library of World Literature, 1962 (originally published in 1826).

Corbett, Julian S. *England in the Seven Years War: A Study in Combined Strategy*. London: Longmans Green, 1918.

Corkran, David H. *The Cherokee Frontier: Conflict and Survival, 1740–62*. Norman: University of Oklahoma Press, 1962.

Cuneo, John R. *Robert Rogers of the Rangers*. New York: Oxford University Press, 1959.

Dorn, Walter L. *Competition for Empire, 1740–1763*. New York: Harper & Brothers, 1940.

Dowd, Gregory Evans. *A Spirited Resistance: The North American Indian Struggle for Unity, 1745–1815*. Baltimore: Johns Hopkins, 1992.

———. *War Under Heaven: Pontiac, the Indian Nations, and the British Empire*. Baltimore: Johns Hopkins, 2002.

Eckert, Allan. *Wilderness Empire*. Boston: Little, Brown and Company, 1969.

Fitch, Jabez, Jr. *The Diary of Jabez Fitch, Jr.* Second edition. Glen Falls, N.Y.: Rogers Island Historical Association, 1968.

Frégault, Guy. Translated by Margaret M. Cameron. *Canada: The War of the Conquest*. Toronto: Oxford University Press, 1969.

Garraty, John A. *The American Nation: A History of the United States*. New York: Harper & Row, 1966.

Gipson, Lawrence Henry, *The British Empire before the American Revolution. Vol. IV, Zones of International Friction: North America, South of the Great Lakes Region, 1748–1754*. New York: Alfred A. Knopf, 1939.

———. *The British Empire before the American Revolution. Vol. V, Zones of International Friction: The Great Lakes Frontier, Canada, the West Indies, India, 1748–1754*. New York: Alfred A. Knopf, 1942.

———. *The British Empire before the American Revolution. Vol. VI, The Great War for the Empire: The Years of Defeat, 1754–1757*. New York: Knopf, 1946.

———. *The British Empire before the American Revolution. Vol. VII, The Great War for the Empire: The Victorious Years, 1758–1760*. New York: Knopf, 1949.

———. *The British Empire before the American Revolution. Vol. VIII, The Great War for the Empire: The Culmination, 1760–1763*. New York: Knopf, 1953.

———. *The British Empire before the American Revolution. Vol. IX, The Triumphant Empire: New Responsibilities within the Enlarged Empire, 1763–1766*. New York: Knopf, 1956.

———. *The Coming of the Revolution, 1763–1775*. New York: Harper and Row, 1962.

Godfrey, William G. *Pursuit of Profit and Preferment in Colonial North America: John Bradstreet's Quest*. Waterloo, Ont.: Wilfrid Laurier University Press: 1982.

Hall, Walter Phelps, and Robert Greenhalgh Albion. *A History of England and the British Empire*. Boston: Ginn and Company, 1953.

Hamilton, Edward P. *Fort Ticonderoga: Key to a Continent*. Boston: Little, Brown, 1964.

Hatley, Tom. *The Dividing Paths: Cherokees and South Carolinians through the Revolutionary Era.* New York: Oxford University Press, 1995.

Hibbert, Christopher. *Wolfe at Quebec: The Man Who Won the French and Indian War.* New York: Cooper Square Press, 1999 (originally published 1959).

Howard, Robert West. *Thundergate: The Forts of Niagara.* Englewood, N.J.: Prentice Hall, 1968.

Isaacson, Walter. *Benjamin Franklin: An American Life.* New York: Simon & Schuster, 2003.

Jacobs, Wilbur R. *Wilderness Politics and Indian Gifts: The Northern Colonial Frontier, 1748–1763.* Lincoln: University of Nebraska Press, 1966.

James, A. P., ed. *The Writings of General John Forbes.* Menasha, Wis.: The Collegiate Press,1938.

Jennings, Francis. *Empire of Fortune: Crowns, Colonies, and Tribes in the Seven Years War in America.* New York: W. W. Norton, 1988.

Johnston, A. J. B. et al. *Louisbourg, An Eighteenth-Century Town.* Halifax: Nimbus Publishing, 1991.

Jomini, Antoine. *The Art of War.* Westport, Conn.: Greenwood Press (originally published by J. B. Lippincott, 1862).

Knollenberg, Bernhard. *Origin of the American Revolution, 1759–1766.* New York: Macmillan, 1960.

Koontz, Louis Knott. *Robert Dinwiddie: His Career in American Colonial Government and Westward Expansion.* Glendale, Calif.: Arthur H. Clark, 1941.

Kopperman, Paul E. *Braddock on the Monongahela.* Pittsburgh: University of Pittsburgh Press, 1977.

Lavender, David. *The Fist in the Wilderness.* Garden City, N.Y.: Doubleday, 1964.

Leach, Douglas Edward. *Arms for Empire: A Military History of the British Colonies in North Amerca, 1607–1763.* New York: Macmillan, 1973.

———. *The Northern Colonial Frontier, 1607–1763.* New York: Holt, Rinehart, and Winston, 1966.

Leckie, Robert. *A Few Acres of Snow: The Saga of the French and Indian Wars.* New York: John Wiley, 1999.

Loescher, Burt Garfield. *Genesis: Rogers' Rangers, The First Green Berets.* Bowie, Md.: Heritage Books, 2000.

———. *The History of Rogers' Rangers: The St. Francis Raid.* Bowie, Md.: Heritage Books, 2002.

Long, J. C. *Lord Jeffery Amherst: A Soldier of the King*. New York: Macmillan, 1933.

———. *Mr. Pitt and America's Birthright: A Biography of William Pitt, the Earl of Chatham 1708–1778*. New York: Frederick A. Stokes, 1940.

Lucier, Armand Francis. *French and Indian War Notices Abstracted from Colonial Newspapers*. Vol. 1–4. Bowie, Md.: Heritage Books, 1999.

Mahan, A. T. *The Influence of Sea Power upon History, 1660–1783*. New York: Dover Publications, 1987 (originally published in 1890).

Mante, Thomas. *The History of the Late War in North America and the Islands of the West-Indies*. London: Strahan and Cadell, 1772.

Marston, Daniel. *The French-Indian War, 1754–1760*. Oxford, UK: Osprey Publishing, 2002.

McCardell, Lee. *Ill-Starred General: Braddock of the Coldstream Guards*. Pittsburgh: University of Pittsburgh Press, 1958.

McLennan, J. S. *Louisbourg from Its Foundation to Its Fall, 1713–1758*. Halifax: The Book Room Limited, 1979.

McLynn, Frank. *1759: The Year Britain Became Master of the World*. New York: Atlantic Monthly Press, 2004.

Miller, John C. *Origins of the American Revolution*. Boston: Little, Brown, 1943.

Moreland, W. H., and Atul Chandra Chatterjee. *A Short History of India*. Fourth edition. New York: David McKay, 1957.

Morison, Samuel Eliot. *The Oxford History of the American People*. New York: Oxford University Press, 1965.

O'Meara, Walter. *Guns at the Forks*. Englewood Cliffs, N. J.: Prentice Hall, 1965.

Pargellis, Stanley McCrory. *Lord Loudoun in North America*. New Haven: Yale University Press, 1933.

Parkman, Francis. *La Salle and the Discovery of the Great West*. In *France and England in North America*, Vol. I. New York: Library of America (Viking Press), 1983.

———. *Montcalm and Wolfe: The French and Indian War*. New York: Da Capo Press, 1995 (originally published in 1884 as volume XX of *France and England in North America*).

———. *The Conspiracy of Pontiac and the Indian War after the Conquest of Canada*. New York: E. P. Dutton, 1908 (reprint).

Pares, Richard. *War and Trade in the West Indies, 1739–1763*. Oxford: Clarendon Press, 1936.

Parry, J. H., Philip Sherlock, and Anthony Maingot. *A Short History of the West Indies*. New York: Macmillan Caribbean, 1987.

Peckham, Howard H. *Pontiac and the Indian Uprising*. Chicago: University of Chicago Press, 1961.

———. *The Colonial Wars: 1689–1762*. Chicago: University of Chicago Press, 1964.

Petrie, Charles. *King Charles III of Spain: An Enlightened Despot*. London: Constable, 1971.

Pratt, Julius W. *A History of United States Foreign Policy*. Englewood Cliffs, N. J.: Prentice Hall, 1955.

Quaife, Milo Milton, ed. *The Siege of Detroit in 1763*. Chicago: Lakeside Press, 1958 (reprint).

Rankin, Hugh F. *Francis Marion: The Swamp Fox*. New York: Thomas Y. Crowell, 1973.

Rashed, Zenab Esmat. *The Peace of Paris*. Liverpool: University Press, 1951.

Reid, Stuart. *Wolfe: The Career of General James Wolfe from Culloden to Quebec*. Rockville Center, N.Y: Sarpedon, 2000.

Richards, Frederick B. *The Black Watch at Ticonderoga and Major Duncan Campbell of Inverawe* (an excerpt from volume X of the Proceedings of the New York State Historical Association printed for the Ticonderoga Museum Library, no date).

Roberts, Kenneth. *Northwest Passage*. New York: Doubleday, 1936.

Robinson, W. Stitt. *The Southern Colonial Frontier, 1607–1763*. Albuquerque: University of New Mexico Press, 1979.

Rogers, Robert. *Reminiscences of the French War with Robert Rogers' Journal and a memoir of General Stark*. Freedom, N.H: Freedom Historical Society, 1988.

Schutz, John A. *William Shirley: King's Governor of Massachusetts*. Chapel Hill: University of North Carolina Press, 1961.

Schwartz, Seymour I. *The French and Indian War, 1754–1763: The Imperial Struggle for North America*. New York: Castle Books, 1994.

Shannon, Timothy J. *Indians and Colonists at the Crossroads of Empire: The Albany Congress of 1754*. Ithaca, N.Y: Cornell University Press, 2000.

Starbuck, David R. *The Great Warpath: British Military Sites from Albany to Crown Point*. Hanover, N.H.: University Press of New England, 1999.

———. *Massacre at Fort William Henry*. Hanover, N.H.: University Press of New England, 2002.

Steele, Ian K. *Betrayals: Fort William Henry and the "Massacre."* New York: Oxford University Press, 1990.

Stille, Samuel Harden. *Ohio Builds a Nation*. Lower Salem, Ohio: Arlendale Book House, 1962.

Sugden, John. *Tecumseh: A Life*. New York: Henry Holt, 1998.

Tebeau, Charlton W. *A History of Florida*. Coral Gables, Fla: University of Miami Press, 1971.

Todish, Timothy J. *America's First World War: The French and Indian War, 1754–1763*. Grand Rapids, Mich.: Suagothel Productions, 1982.

Voltaire. *Candide*. New York: Random House, 1930 (Modern Library reprint).

Wallbank, T. Walter., et al. *Civilization: Past and Present*. Glenview, Ill: Scott, Foresman and Company, 1967.

Ward, Matthew C. *Breaking the Backcountry: The Seven Years' War in Virginia and Pennylsvania, 1754–1765*. Pittsburgh: University of Pittsburgh Press, 2003.

Willcox, William B. *The Age of Aristocracy, 1688–1830*. Volume 3 in *A History of England*. Lexington, Mass.: D. C. Heath and Company, 1971.

Wilson, Beckles. *The Life and Letters of James Wolfe*. London: William Heinemann, 1909.

Articles

Clayton, T. R. "The Duke of Newcastle, the Earl of Halifax, and the American Origins of the Seven Years' War." *Historical Journal* 24 (1981), 571–603.

Dorn, Walter L. "Frederic the Great and Lord Bute." *Journal of Modern History*, 1:4 (December 1929), 529–560.

Harkness, Albert Jr. "Americanism and Jenkins' Ear." *Mississippi Valley Historical Review* 37 (1950), 61–90.

Humphrey, Helen F. "The Identity of Gladwin's Informant." *Mississippi Valley Historical Review* 21:2 (September 1934), 147–162.

Jacobs, Wilbur R. "Was the Pontiac Uprising a Conspiracy? *Ohio State Archaeological and Historical Quarterly* 59:1 (January 1959), 26–37.

Koontz, Louis Knott, "The Virginia Frontier, 1754–1763." *Johns Hopkins University Studies in Historical and Political Science* 43 (Baltimore: Johns Hopkins Press, 1925).

Lambing, A. A., ed. "Account of the Voyage on the Beautiful River Made in 1749, under the Direction of Monsieur De Céloron, by Father Bonnecamps." *Ohio State Archaeological and Historical Quarterly* 29:3 (July 1920), 397–423.

———. "Céloron's Journal." *Ohio State Archaeological and Historical Quarterly* 29:3, (July 1920), 335–396.

Marshall, O. H. "De Céloron's Exposition to the Ohio in 1749." *Ohio State Archaeological and Historical Quarterly* 29:3 (July 1920), 424–450.

Pargellis, Stanley. "Braddock's Defeat." *American Historical Review* 41 (1936), 253–269.

Smelser, Marshall. "The Insular Campaign of 1759: Guadeloupe." *American Neptune* 7:1 (January 1947), 21–34.

———. "The Insular Campaign of 1759: Martinique." *American Neptune* 6:4 (October 1946), 290–300.

Papers and Government Documents

Adams, John. *Diary and Autobiography of John Adams*. Ed. L. H. Butterfield. Vol. 1: Diary 1755–1770. Cambridge, Mass.: Belknap Press of Harvard University Press, 1961.

Amherst, Jeffery. *The Journal of Jeffery Amherst: Recording the Military Career of General Amherst in America from 1758 to 1763*. Ed. J. Clarence Webster. Toronto: Ryerson Press, 1931.

Bougainville, Louis Antoine de. *Adventure in the Wilderness: The American Journals of Louis Antoine de Bougainville, 1756–1760*. Ed. Edward P. Hamilton. Norman: University of Oklahoma Press, 1964.

Bouquet, Henry *The Papers of Henry Bouquet*. Eds. S. K. Stevens, Donald H. Kent, Autumn L. Leonard. Vol. 1: December 11, 1755–May 31, 1758. Harrisburg: The Pennsylvania Historical and Museum Commission, 1972.

———. *The Papers of Henry Bouquet*. Ed. S. K. Stevens, Donald H. Kent, Autumn L. Leonard. Vol. 2: The Forbes Expedition. Harrisburg: The Pennsylvania Historical and Museum Commission, 1951.

Boyd, Julian P. *The Susquehannah Company Papers*. Vol. 1: 1750–1755. Ithaca, NY: Cornell University Press, 1962.

Documents Relating to the Colonial History of the State of New York. Albany, N.Y. : Weed, Parsons, 1855.

Forbes, John. *Writings of General John Forbes*. Ed. Alfred Proctor James. Menasha, Wis.: The Collegiate Press, 1938.

Franklin, Benjamin. *The Papers of Benjamin Franklin*. Ed. Leonard W. Larabee. Vol. 4, July 1, 1750, through June 30, 1753. New Haven, Conn: Yale University Press, 1961.

———. *The Papers of Benjamin Franklin*. Ed. Leonard W. Larabee. Vol.

5, July 1, 1753, through March 31, 1755. New Haven, Conn.: Yale University Press, 1962.

———. *The Papers of Benjamin Franklin*. Ed. Leonard W. Larabee. Vol. 10: January 1, 1762, through December 31, 1763. New Haven, Conn.: Yale University Press, 1966.

———. *The Writings of Benjamin Franklin*. Ed. Albert Henry Smyth. Vol. 3: 1750–1759. New York: Macmillan, 1907.

Johnson, William. *The Papers of Sir William Johnson*. Ed. James Sullivan. Albany: The University of the State of New York, 1921.

Pennsylvania Colonial Records. Volumes 4, 5, and 6.

Pitt, William. *Correspondence of William Pitt When Secretary of State with Colonial Governors and Military and Naval Commissioners in America*. Ed. Gertrude Selwyn Kimball. New York: Macmillan, 1906.

Shirley, William. *Correspondence of William Shirley*. Ed. Charles Henry Lincoln. New York: Macmillan, 1912.

Washington, George. *The Journal of George Washington: An Account of His First Official Mission, Made as Emissary from the Governor of Virginia to the Commandant of the French Forces on the Ohio, October 1753–January 1754*. Williamsburg, Va.: Colonial Williamsburg, 1959 (facsimile reprint of 1754 Hunter edition).

———. *The Papers of George Washington*. Ed. W. W. Abbot. Colonial Series, vol. 1, 1748–August 1755. Charlottesville: University Press of Virginia, 1983.

———. *The Papers of George Washington*. Ed. W. W. Abbot. Colonial Series, vol. 5, October 1757–September 1758. Charlottesville: University Press of Virginia, 1988.

———. *The Papers of George Washington*. Ed. W. W. Abbot. Colonial Series, vol. 6, September 1758–December 1760. Charlottesville: University Press of Virginia, 1988.

Newspapers

Boston Evening Post

Boston Gazette

Halifax Gazette

London Evening Post

Maryland Gazette (Annapolis)

New York Gazette

New York Mercury

Pennsylvania Gazette (Philadelphia)

South Carolina Gazette (Charleston)

INDEX

About the Author

About the Book

Insights,
Interviews
& More ...

Read On

Meet
Walter R. Borneman

The author at Fort Ticonderoga

WALTER R. BORNEMAN is an American historian who has written books and articles about mountains, railroads, and America's westward expansion, including *1812: The War That Forged a Nation* (HarperCollins, 2004) and *Alaska: Saga of a Bold Land* (HarperCollins, 2003). In his home state of Colorado, he is best known as the coauthor of *A Climbing Guide to Colorado's Fourteeners*, first published in 1978 and in print for nearly thirty years. Borneman has worked for the Colorado Historical Society; practiced law, with frequent involvement in historic preservation issues; and served as the first chairman of the Colorado Fourteeners Initiative, a nonprofit organization devoted to the preservation of the state's highest peaks. He is the president of the Walter V. and Idun Y. Berry Foundation, which funds postdoctoral fellowships in children's health at Stanford University.

For more information about Walter R. Borneman and his books, please visit walterborneman.com. ❧

" [Borneman] served as the first chairman of the Colorado Fourteeners Initiative, a nonprofit organization devoted to the preservation of the state's highest peaks. "

Historian on the Road

"STOP!" MARLENE SCREAMED. "There's a canoe!"
We were driving along the waters of Lake
George in upstate New York and I was deep
in thought about Major Rogers and his rangers
leading Abercromby's grand flotilla down the
lake toward Fort Carillon. But my wife's eyes
were not fixed on the lake. She had caught a
glimpse of a miniature canoe in a shop window
in the delightful hamlet of Bolton Landing.
Knowing very well that history is full of detours,
I steered into the next available parking space
and we hustled back for a closer look.

I rarely make it to *all* the places in my
books, but it's not for lack of trying. A good
grasp of geography and the lay of the land
is essential to understanding events. And,
certainly, being on the road visiting historic
sites and walking the ground that I write
about has always been my favorite part
of the research and writing process.

Some places have changed beyond
recognition. At other locations, such as Fort
Necessity National Battlefield, it's possible
to walk through woods and meadows and
easily imagine the rattle of muskets and
the smell of black powder. Occasional
reenactments there and elsewhere add
even more realism.

Lake George and upper Lake Champlain
are at the heart of French and Indian War
country. One cornerstone is Fort William
Henry at the southern tip of Lake George.
This reconstruction and its living history
program, where children "enlist" either in
the King's army or as Native American
allies, is one of the best of its kind.

North of Lake George at Fort
Ticonderoga, the main fort with its
Revolutionary War additions gets the
attention, but for the French and Indian ▶

> ❝ Being on
> the road visiting
> historic sites
> and walking the
> ground that I
> write about has
> always been my
> favorite part
> of the research
> and writing
> process. ❞

Fort William Henry

Historian on the Road *(continued)*

War period, it is most interesting to walk the woods on the promontory to the west. Here, Abercromby's splendid lines were ground up in the face of French trenches and abatis. Looking back toward Mount Defiance, one can ponder what British artillery on those heights might have done instead.

At Crown Point, one can see the ruins of Fort Saint Frédéric—its size hardly seems to qualify it as one of New France's imposing colonial bastions—as well as the huge earthworks remaining from Jeffery Amherst's later exercise in fort construction. Looking out across the waters of Lake Champlain, it's easy to see why Amherst pronounced the location "the finest situation" that he had seen in North America.

Those who profess history to be boring simply haven't taken the time to experience it outside of the pages of some textbook. Walk the ground; experience the locations; ponder the feelings of the participants. Only then can you begin to understand and appreciate the broader context of an event.

Oh, what about that canoe? At around twenty-four inches in length and complete with carved paddles and tiny little baskets for gear, it proved to be the perfect souvenir from our travels to research *The French and Indian War*. 〜

> ❝ Those who profess history to be boring simply haven't taken the time to experience it outside of the pages of some textbook. ❞

4

Women in the French and Indian War

THOUGH WOMEN are seldom mentioned in either contemporary or secondary accounts, it should not be assumed that they played little or no role in the French and Indian War. On all three sides—British, French, and Native American—women were an integral, if in hindsight often neglected, part of the drama.

On the British side, colonial women fought and suffered alongside the men, although newspaper reports of women being killed, scalped, or captured frequently listed only that the victim was "the Widow Jones" or someone's wife, rather than giving a first name.

In the spring of 1756, one account of a skirmish in central Pennsylvania reported that "Mary McCord was shot by the fire of our own men, and that the wife of John Thorn, with a child at her breast, two of William McCord's daughters, and a little boy, made their daring escape during the engagement and are safe at Fort Littleton."

Later that year, a British raid at Kittanning, a town northeast of Fort Duquesne, resulted in the recovery of a number of women captured earlier. They were listed in a Pennsylvania newspaper along with the location where they had been taken: "Anne McCord, at Conegocheague. Martha Thorn, about seven years old, taken at the same place. Barbara Hicks, taken at the Connoloways. Catherine Smith, a German child, taken near Shamokin, Margaret Hood taken near the mouth of Conegocheague in Maryland . . . and Sarah Kelly taken near Winchester." [Lucier, *Newspapers*, Vol. II, pp. 55, 139.]

The twenty-to-one advantage in general population that the British held over the French in North America was even more ▶

> **❝** Newspaper reports of women being killed, scalped, or captured frequently listed only that the victim was 'the Widow Jones' or someone's wife, rather than giving a first name. **❞**

5

Women in the French and Indian War
(*continued*)

lopsided among females. In large part, this was because of the basic difference in colonization strategies: New France was characterized by the transitory trader, the English colonies by the farmer pushing his plow with his brood close behind. Needless to say, the smaller percentage of French females did nothing to close the already disparate population numbers.

The French government made some effort to bolster the female population in Canada and Louisiana by exporting shiploads of young women from France. Sometimes called *filles à la cassette* (casket girls), they arrived with all of their worldly belongings in a single trunk resembling a small casket. Unfortunately, many met an early death from childbirth, disease, and the general rigors of the frontier.

The shortage of women in New France led to many marriages and liaisons between Frenchmen and Native Americans. This resulted in a growing number of métis offspring, who were sometimes caught between two cultures, particularly after the French exodus. In the traditional matriarchal system of most tribes—where descent and property rights were determined through the female line—Native American women usually had much to say about these family arrangements, as well as the "property rights" to captives.

Certainly, women were no strangers on the battlefields of the war. Women and children marched with the British column that was attacked after the surrender of Fort William Henry. One diarist who was close by at Fort Edward wrote of children being taken from their mothers and beaten to death. [Fitch, *Diary*, p. 19.]

Although ultimately not very successful, General Edward Braddock's early efforts to

> ❝ The shortage of women in New France led to many marriages and liaisons between Frenchmen and Native Americans. ❞

recruit Native American allies resulted in some warriors arriving in camp accompanied by their wives. When Braddock began his own march, forty to fifty British women were reportedly in his advance column. Most were wives—common-law or otherwise—of soldiers or teamsters who performed a variety of washing and cooking chores. A few appear to have been the proverbial "camp followers."

Contrary to oft-repeated rumors, the most careful scholar of the Monongahela battle, Paul Kopperman, has found no evidence to support the story that Braddock had a mistress among the women in the column. Other reports claim that the bodies of eight women were among the dead and that at least seven women were captured. [Kopperman, *Braddock at the Monongahela*, pp. 31, 100, 137; O'Meara, *Guns at the Forks*, pp. 148, 150.]

If a letter from Albany published in a Boston newspaper is accurate, one of the women in Braddock's column ended up with what might have been the most unusual experience of the war:

> When the French prisoners lately taken at Niagara, arrived in this city, on their way down hither, an English woman, wife of one of the soldiers that was in General Braddock's army, having been taken prisoner by the French at the time of the defeat of General Braddock, and supposing that her husband was slain at that time, during her imprisonment married a French subaltern, by whom she has had one child. Being with her [French] husband coming prisoner through Albany, she was there discovered by her former husband, who was then on duty there. He immediately demanded her, and after some struggles of tenderness for her ▶

> ❝ One of the women in Braddock's column ended up with what might have been the most unusual experience of the war. . . . ❞

Women in the French and Indian War
(continued)

French husband, she left him, and closed again with her first; though 'tis said the French husband insisted on keeping the child, as his property, which was consented to by the wife and her first husband. [Lucier, *Newspapers*, Vol. III, p. 348.] ∽

More About the Native American "What If"

ONE OF THE INTRIGUING "what ifs" of the French and Indian War and its aftermath is the possibility of a Native American confederacy that would have blocked colonial expansion west of the Appalachians—at least temporarily. The British government in London provided support for this notion—consciously or not—with the Proclamation of 1763, which professed to create a vast "Indian Reserve."

Colonials on the ground in North America ignored this decree, of course, just as they had ignored French claims to dominion over the Ohio country before the war. But that did not mean that Native Americans were willing to accept escalating colonial encroachments without a fight, or that their resistance was only sporadic and localized.

Pulitzer Prize–winning historian Alan Taylor (*William Cooper's Town* and *The Divided Ground*) has argued convincingly that prior to the War of 1812, the Anglo triumph over Native Americans was far from inevitable. At the close of the French and Indian War, the Six Nations wielded considerable influence over the territory that Taylor calls "Iroquoia," an immense borderland separating southern New York from Quebec and Upper Canada. Only as this land—Taylor's "divided ground"— was broken down over a generation of conflict and land manipulations did the power of the Iroquois wane.

Later historians have generally recognized that, west of Lakes Erie and Ontario, Pontiac was not the seminal architect of a conspiracy as unified and comprehensive as Francis Parkman professed it to be. Contemporaneous uprisings from Fort Michilimackinac to Fort Niagara were less evidence of Pontiac's organizational genius than they were of the level of mutual ▶

More About the Native American "What If"
(continued)

discord among Native Americans and of their universal determination to seize upon the French exit as the moment to resist new British incursions.

But what if the forts at Detroit, Niagara, and Pitt had not held out against these Native American attacks? One obvious result would have been that the frontier of the thirteen colonies would have been pushed back to the Appalachians. This might have encouraged additional dependence on the British Crown for protection and blunted rumblings for independence. It might also, of course, have reignited the cries for intercolonial cooperation for self-defense that Franklin and others had championed at Albany in 1754.

Once pushed back, the colonial frontier might have been dammed along the Appalachians indefinitely had an Ohio Indian confederation similar to the Six Nations taken shape. This might have opposed any reentry into the Ohio River country or Daniel Boone's later migration through the Cumberland Gap. Taken to the extreme, such a confederation extending southward to embrace the Cherokee and Creek nations would have completed the encirclement of the thirteen colonies. Unlikely? In retrospect, perhaps, but Tecumseh was still trying to accomplish this sort of concerted opposition just prior to the War of 1812.

Another possible scenario is that a vigorous confederation, or allied confederations, might have formed a strong trading relationship with the colonies, while at the same time exerting more formal control over their lands and avenues of ingress. Imagine colonial representatives gathering in Albany or elsewhere to renew the Covenant Chain, and the Iroquois—perhaps because of a British trade embargo against its wayward

66 What if the forts at Detroit, Niagara, and Pitt had not held out against these Native American attacks? 99

colonies—holding the undisputed upper hand.

Of course, this is all conjecture. The forts at Detroit, Niagara, and Pitt held, and by the time that Henry Bouquet and John Bradstreet marched west on punitive missions in 1764, coordinated resistance among the Ohio Indians was minimal. In large part because of their stronger record of confederation, the Six Nations remained a more potent force until a burgeoning colonial population made steady inroads. Those interested in why this happened sooner rather than later should consult Professor Taylor's books as well as other recent scholarship on Native American attempts at confederation. ∽

Why Didn't Quebec Become the Fourteenth Colony?

CANADA IS OFTEN termed the closest ally that the United States has ever had. But that doesn't mean that this was always the case. With France expelled from most of North America, the question might well be asked, why didn't Quebec join the thirteen colonies in revolt against Great Britain?

The answer lies both in the strong French heritage of the region and in Great Britain's response to it. The hoisting of the Union Jack over Montreal or the scratch of a pen on the Treaty of Paris could not change the fact that, for a century and a half, French language, laws, and culture were ingrained among the inhabitants of the Saint Lawrence River valley. After 1763, rather than aggressively assaulting these institutions, the British indirectly encouraged them—frequently to the chagrin of those living in the thirteen colonies.

The most blatant example of this was the Quebec Act of 1774, which not only went to great lengths to appease French Canadians by recognizing the role of the Roman Catholic Church, but also extended Quebec's boundaries over lands previously claimed by other colonies. Most egregious to New York, Pennsylvania, and Virginia expansionists was the extension of Quebec's boundaries to include lands they had long coveted north of the Ohio River.

Then too, while legally unified under one flag, Quebec was still well isolated from the other colonies by a rugged frontier—even if not by a monolithic Native American nation. It was much more difficult to travel to New York from Quebec or Montreal than from Boston or Baltimore. Isolated geographically and politically from their new cousins and mollified by a government in London that seemed content not to ruffle tempers along the Saint Lawrence, Quebec residents found no burning reason to rise in revolt. And the longer they refrained from doing so, the more determined they became to hold American entreaties at arm's length.

Not only did Quebec refuse to become the fourteenth colony, but it resisted almost a century of American attempts to acquire it and all of Canada. In 1775, an American invasion led by Richard Montgomery and Benedict Arnold was repulsed in wintry weather from the very gates of the city that had surrendered to James Wolfe. During the War of 1812, an emerging Canadian identity stood firm on the side of British regulars and defeated three years of multipronged American offensives. Even after Canada achieved some measure of self-government in 1867, there were still Americans who demanded that Great Britain cede all of Canada to the United States as reparation for British aid to Confederate raiders during the Civil War.

> Not only did Quebec refuse to become the fourteenth colony, but it resisted almost a century of American attempts to acquire it and all of Canada.

...And Don't Forget the Ongoing 250th Anniversary Commemoration

A VARIETY of celebrations, reenactments, and exhibits are under way to commemorate the 250th anniversary of the French and Indian War. Begun in North America in 1754 with Washington's defeat at Fort Necessity, the war became global two years later when Great Britain and France issued formal declarations and stepped up military efforts. What Europe came to call the Seven Years' War ended in 1763 with the Treaty of Paris; 2013 will mark the 250th anniversary of Pontiac's siege of Detroit as well as other Native American offensives.

Check with individual historic sites for a schedule of events or consult the calendar maintained by French and Indian War 250 at www.frenchandindianwar250.org. In particular, Fort Ticonderoga and Old Fort Niagara are planning major events in 2008 and 2009, respectively. ᕙ

More from
Walter R. Borneman

1812: THE WAR THAT FORGED A NATION

Although frequently overlooked between the
American Revolution and the Civil War, the
War of 1812 tested a rising generation of
American leaders, unified the United States
with a renewed sense of national purpose,
and set the stage for westward expansion
from Mackinac Island to the Gulf of
Mexico. The USS *Constitution*, "Old
Ironsides," proved the mettle of the
fledgling American navy; Oliver Hazard
Perry hoisted a flag boasting "Don't
Give Up the Ship"; and Andrew Jackson's
ragged force stood behind its cotton
bales at New Orleans and bested the
pride of British regulars. Here are the
stories of commanding generals such
as America's double-dealing James
Wilkinson and Great Britain's gallant
Sir Isaac Brock, of Canada's heroine farm
wife Laura Secord, and of country doctor
William Beanes, whose capture set the
stage for Francis Scott Key to write
"The Star-Spangled Banner." During the
War of 1812, the United States cast off its
cloak of colonial adolescence and—with both
humiliating and glorious moments—found the
fire that was to forge a nation.

"Without question, this is the best popular
account of the War of 1812." —Robert V. Remini

"A thoroughly readable popular history of
the War of 1812 . . . strong in vivid personal
portraits (the gigantic Winfield Scott and the
diminutive and sickly James Madison) and
evenhanded as far as atrocities (too many, by
all parties) are concerned." —*Publishers Weekly*

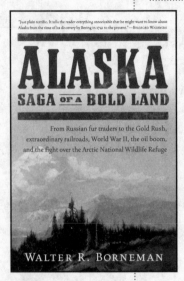

The history of Alaska is filled with stories of new land and new riches—and ever present are new people with competing views over how the valuable resources should be used: Russians exploiting a fur empire; explorers checking rival advances; prospectors stampeding to the clarion call of "Gold!"; soldiers battling out a decisive chapter in world war; oil wildcatters looking for a different kind of mineral wealth. Always at the core of these disputes is the question of how the land is to be used and by whom.

While some want Alaska to remain static, others are in the vanguard of change. *Alaska: Saga of a Bold Land* shows that there are no easy answers on either side and that Alaska will always be crossing the next frontier.

"This is narrative history told in superlatives."
  —David Lavender, author of *The Great West*

"The balance of textbook history and storytelling . . . makes this informative book so readable."   —*Publishers Weekly*

The Web Detective

*Check out the following Web sites for
more information about the locations
and events of the French and Indian War.*

Crown Point State Historic Site
http://www.nysparks.state.ny.us/sites

Fort Ligonier
http://www.fortligonier.org

Fort Necessity National Battlefield
http://www.nps.gov/fone

Fort Loudoun
http://www.fortloudoun.com

Fort Ontario
http://www.fortontario.com

Fort Pitt Museum
http://www.fortpittmuseum.com

Fort Stanwix National Monument
http://www.nps.gov/fost

Fort Ticonderoga
http://www.fort-ticonderoga.org

The Fort at No. 4 Living History Museum
http://www.fortat4.com

Fortifications of Québec National Historic Site
(Canada)
http://www.parkscanada.ca

Fortress of Louisbourg National Historic Park
(Canada)
http://www.louisbourg.ca/fort

The Web Detective *(continued)*

John Heinz Regional History Center
http://www.pghhistory.org

Old Fort Niagara
http://www.oldfortniagara.org

Don't miss the next
book by your favorite
author. Sign up now for
AuthorTracker by visiting
www.AuthorTracker.com.